**Northamptonshire
County Council**
Libraries and Information S~·

42

KALASHNIKOV

GREENHILL MILITARY MANUALS

Greenhill Books

To ARW and ADW
for their patience

KALASHNIKOV

MACHINE PISTOLS, ASSAULT RIFLES AND MACHINE-GUNS, 1945 TO THE PRESENT

GREENHILL MILITARY MANUALS

John Walter

Greenhill Books, London
Stackpole Books, Pennsylvania

Greenhill Books

Kalashnikov

first published 1999 by Greenhill Books, Lionel Leventhal Limited, Park House, 1 Russell Gardens, London NW11 9NN

and

Stackpole Books, 5067 Ritter Road, Mechanicsburg, PA 17055, USA

© John Walter, 1999

The moral right of the author has been asserted

British Library Cataloguing in Publication Data

Walter, John, 1951–
Kalashnikov : machine pistols, assault rifles and machine-guns, 1945 to the present. – (Greenhill military manuals)
1.Kalashnikov machine gun 2.Kalashnikov machine gun – History
I.Title
623.4'424

ISBN 1-85367-364-1

Library of Congress Cataloging-in-Publication Data

Walter, John.
Kalashnikov : machine pistols, assault rifles and machine-guns, 1945 to the present. / by John Walter.
p.   cm. -- (Greenhill military manuals)
ISBN 1-85367-364-1
1. AK-47 rifle. I. Title. II. Series.
UD395.A16W35 1999
623.4'424--dc21
98-46536 CIP

Printed and bound in Singapore

Contents

Foreword

The history of Russian small arms can be greatly affected by the state of Russian politics at the time that it is being written. The status of inventors can fluctuate, and there is an unfortunate tendency to claim originality when it is much more likely that captured weapons had simply been copied. For example, the 7.62x39 M43 'intermediate' cartridge was most probably based on the German 7.9mm Kurz pattern, and Russian claims that research began in pre-war days — while not actually incorrect — are very misleading.

Yet, paradoxically, there are areas in which the Russians have led other European arms manufacturers. The most obvious of these is the mass production of the Tokarev semi-automatic rifle, at a time when even the Germans were still struggling with clumsy and inefficient Bang-type muzzle-trap guns. It was ironic that the Tokarev failed more because of the poor standards of training prevalent in the Red Army than because of its inherent operating problems.

Undoubtedly, the success of the Kalashnikov rifle in post-war days has obscured its true genesis. Its gas system clearly owes something to the Tokarev rifle, the trigger is too similar to the Garand to be entirely coincidental, while the concept of a rotating-bolt lock had existed for many years before the rifle appeared. However, the Russians did succeed in combining into a battleworthy weapon the ideas that the West struggled to duplicate until the 1960s.

The Kalashnikov has served as the first choice not only of regular forces with pro-Communist leanings, but also of countless terrorist groups operating from South America to the Far East. It is ironic, therefore, that the peaceful political evolution in Europe during the 1990s has finally thrown into question the future of the Kalashnikov — arbiter of much more violent change. It will be irresting to see, having reached its half-century, whether the gun will be looked upon merely as a twentieth-century icon: even in Russia, the current 'Hundred Series' Kalashnikov assault rifles are under threat from the Nikonov prototypes.

The production of at least seventy million Kalashnikov and Kalashnikov-type assault rifles has made the inventor's name common currency in English, alongside a mere handful of other guns and gunmakers. Much of the story is still to be written, but hopefully this small book will provide a reliable guide to the many different weapons derived directly from the *Avtomat Kalashnikova* of 1947.

I would like to acknowledge the assistance of Ian Hogg, who very kindly supplied many of the photographs. Many thousands of words have been written about the Kalashnikov over the last thirty years; *Sovetskoye strelkovoe oruzhiye za 50 let* ('Soviet infantry weapons of the last 50 years', 1983) and *Soviet Small-Arms and Ammunition* (1995), by David N. Bolotin, and *The AK47 Story* (1986) by the late Edward C. Ezell, have been particularly useful. And Alison and Adam — dare I call them long-suffering? — have accepted Kalashnikovs under the stairs or rooms full of paperwork with good humour, allowing me to complete work under no more than the usual pressure.

John Walter
1999

INTRODUCTION

When World War Two ended in May 1945, the Soviet Army was the most powerful of all land forces, having more men under arms than even the USA. Undoubtedly, its rejuvenation had been helped by Lend-Lease, but the restoration of industries shattered by the German invasion in 1941 — in so short a time — had been a genuine triumph. The quality of Soviet armoured vehicles and aircraft had improved greatly, although small arms design still lagged some way behind Britain, Germany and the United States. This deficiency concerned manufacturing standards rather than basic operating principles, however, and it is true to say that the 1945-vintage Russian small arms worked well enough to satisfy Soviet Army tactics.

There is evidence to show that the Russians habitually accepted new and untried designs that would have failed most Western adoption trials, but reliance on the doctrine of mass-attack — with its attendant high casualty figures — camouflaged comparatively high failure rates. That a Russian submachine-gun jammed once every twenty shots, compared with one in fifty for its German equivalent, was irrelevant; neither did it matter, when

The German MP. 43 — derived from the Haenel prototype shown here — was the first assault rifle to be chambered successfully for an 'intermediate' cartridge. Undoubtedly, guns captured by the Russians provided the inspiration for the Sudaev and Kalashnikov designs.

Soviet troops carrying Tokarev rifles, with knife bayonets fixed, parade in Red Square to mark the 31st anniversary of the 1917 revolution in October 1948.

three times as many guns could be fielded, that Russian light machine-guns broke components more regularly than their technically superior German equivalents. Simplicity, at the expense of sophistication, became the byword of Soviet small-arms design.

By 1943, the Russian Army had captured examples of the first German *Maschinenkarabiner* (machine carbines) on the Eastern Front. The concept of these early assault rifles intrigued Russian designers, who had experimented in the late 1930s with similar guns chambering enlarged pistol/submachine-gun cartridges.

Once the German gun and its 7.9mm Kurz cartridge had been analysed, the merits of such equipment in the hands of the *tankoviy desant* troops (tank-borne infantry) became obvious: the assault rifle combined the fire rate of a submachine-gun with a cartridge that offered perhaps eight times the maximum effective range of the standard 7.62x25 type. Design of the Russian 7.62x39 M43 (or obr. 1943g) cartridge was speedily completed; by August 1944, Major-General Talakin, head of the Artillery Operations Department of the Chief Artillery Directorate, was reporting to the CAD Director, Lieutenant-General Chechulin, that 'a need is being noted...for increases in the combat efficiency of submachine-guns and an extension of effective range...up to approximately 500 metres, with accuracy correspondingly improved.'

The SKS is a short rifle chambered for the 7.62mm M43 'intermediate' cartridge. This is a Yugoslavian M59/66 version, with an integral grenade launcher, but lacking the folding knife-bladed bayonet beneath the muzzle.

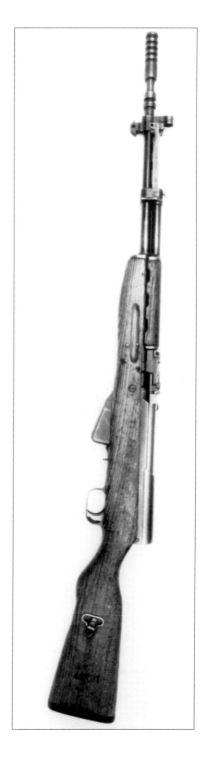

Earlier in 1944, Aleksey Sudaev, best known for the PPS submachine-gun, had submitted a prototype blowback assault rifle, or *avtomat*. The weight of its bolt had been increased considerably to allow for the greater power of the 7.62x39 cartridge; the firing pin had become an integral part of the bolt in pursuit of simplicity; and the trigger mechanism allowed a choice of single shots or automatic fire. The selector lay on the left side of the receiver, while a safety catch ran laterally through the receiver above the pistol grip.

The Sudaev avtomat had a wooden butt with a separate pistol grip, a light stamped-sheet bipod beneath the muzzle, and a detachable, staggered-row, thirty-five-round box magazine ahead of the trigger guard. The tangent-type back sight was graduated to 800m. A knife bayonet could be attached by means of a lug beneath the front-sight block and a crossguard ring that slipped over the muzzle.

Testing revealed that the prototype assault rifle worked well enough, but that the firing pin and the ejector were too weak to withstand rigorous service. Before any remedial work could be undertaken, Sudaev presented a new design in August 1944.

Although externally similar to its predecessor, the new assault rifle was operated by a conventional gas-piston arrangement in a tube above the barrel. It was clear that the layout had been influenced by the SVT, or Tokarev, rifle. However, unlike the latter, the Sudaev bolt was displaced laterally into the receiver wall instead of downward into the receiver. The hammer-type trigger mechanism could be set for single shots or fully automatic fire, a radial-lever selector being provided on the right-hand side of the receiver. A safety catch (inside the trigger guard) could be applied to block the hammer.

The magazine, sights, bipod and bayonet were similar to the earlier blowback-type rifle, but the cartridge

13

capacity had been reduced to thirty rounds, and the bayonet attachment ring slipped over the reduced-diameter tip of the compensator instead of the muzzle. A wooden stock had been retained, and there was a separate wooden hand guard above the barrel.

The testers concluded that the rifle operated reliably and displayed few obvious weaknesses, so a batch of guns — perhaps fifty — was delivered in 1945 for field trials. Unfortunately, the Sudaev assault rifle was much heavier than the submachine-guns in current use and proved too cumbersome to impress the rank-and-file of the Soviet Army. As a result, the Artillery Committee of the Chief Artillery Directorate ordered work to begin on a lightweight version, but Sudaev's unexpected death in 1946 (aged only 34) brought work to an abrupt end.

A temporary solution seems to have been found in the 7.62x54 SKS-41 (Simonov) auto-loading carbine, which had shown great promise prior to the German invasion of Russia. This was successfully adapted to chamber the new 'intermediate' cartridge, proving accurate and reliable once a few teething troubles had been overcome. Eventually, it was adopted as the SKS-45.

Simonov's carbine, however, was a comparatively conventional gun and could not be fired fully automatically. Consequently, the hunt for an assault rifle continued until a solution was found in the prototype submitted in 1946 by Mikhail Kalashnikov, then little known outside restricted circles within the armed forces.

Although some recent biographies have painted Kalashnikov's rise to prominence as a gun designer as a triumph of self-taught skill over formal education — and Kalashnikov himself is apt to labour the point — it should be noted that he had undergone basic engineering training in the Alma-Ata railway workshops before joining the army. Moreover, his design for battle tank testing equipment had been accepted for service prior to the war. However, even his first assault rifle design overcame many of the problems that had taxed much more experienced Soviet designers in their attempts to produce effective guns of this type.

Kalashnikov subsequently recalled that he had wanted his new assault rifle 'to be reliable, compact, light and simple... I could use the blowback principle employed by the existing PPSh and PPS; then the design would be simple. However, the new cartridge for which the assault rifle was created made the bolt too massive, which led to increases in the size and weight of the entire weapon. Additional problems were caused by the new cartridges, which were much longer than the pistol ones. So I decided in favour of the gas piston.

'This scheme allowed a light, portable and reliable weapon with a good rate of fire. Gradually, the outlines of the assault rifle started appearing on paper... However, even the most insignificant change in the shape or size of an individual component forced me to correct all the drawings that had already been made. At last, the design was ready. "What will the experts say?" I thought, waiting for the reply from Moscow. Soon the letter came, telling me that my design had been considered satisfactory and that a prototype could be made. Work started again.'

Readied in Alma-Ata in 1946, with assistance from a designers' collective, the original, or 1946-pattern, Kalashnikov prototype displayed the general lines of the subsequent perfected AK, even though it differed considerably in detail. The butt was attached to a sheet-steel shoe that had been riveted to the receiver; the rear surface of the breech cover was noticeably squared; and the charging handle, safety catch and selector lever were all found on the left side. Thus, the gun could be charged without disturbing the trigger hand, although the entire top surface of the breech was exposed as the bolt ran back.

Thereafter, Kalashnikov spent much of his time working in the Kovrov machine-gun factory, a few hundred kilometres from Moscow. Aleksandr Zaytsev subsequently recalled how, in the

Alone among the countries that became surrounded by the Iron Curtain, Czechoslovakia adopted small arms of indigenous design. The vz. 52 semi-automatic rifle clearly displays the influences of the SKS in the general outline of its receiver and the attachment of a folding bayonet, but is substantially different internally.

autumn of 1946, he had first met Mikhail Kalashnikov: 'Having shown me his 7.62mm carbine for the 1943 cartridge and the design of his assault rifle, Mikhail Timofeevich suggested that I check the technical aspects and prepare the specifications needed to manufacture a prototype of the 7.62mm assault rifle for tests in the factory. Subsequently, after the specifications had been improved on the basis of the test results, two more samples were to be made for firing-range testing. All that had to be done before the end of 1946.

'Assembly began in November. Our models received the code names of AK-1

and AK-2. The tests on the spot were completed by fitter B.P. Marinychev and Kalashnikov himself. Soon two more assault rifles were manufactured.

'At that time, I was engaged in preparing technical documents for the firing range, which took me a great deal of time. By the end of 1946, the difficulties had all been overcome; the prototypes had been shipped to the firing ground, where Mikhail Timofeevich had already gone. And although I had already been given another assignment, I could not help thinking about the ways in which the AK-1 could still be revised. So I started sketching improvements. In the course of the firing-

ground tests, the Kalashnikov assault rifle performed well enough to be admitted to the second stage. The competitors were designers A.A. Demetev from Kovrov and F. Bulkin from Tula, but their guns jammed more frequently under both normal and difficult firing conditions.'

Zaytsev claimed to have suggested extensive changes to the assault rifle in the face of strong objections from Kalashnikov, who was worried that redesigning components was too risky in the middle of testing. Alterations were made, however, and the perfected weapon was submitted to the authorities. Tests on the firing range soon showed that the risks had been worth

The Czechoslovakian vz. 58 assault rifle, while sharing the same general lines as the Kalashnikov, is quite different internally. The breech is locked by a tipping block instead of a rotating bolt.

16 *The Czech vz. 58T assault rifle has a butt that folds back along the right side of the receiver, reducing overall length considerably.*

Although the Soviet Union had settled on an 'intermediate' cartridge for the Kalashnikov (its principal infantry weapon), NATO and the USA remained firmly wedded to heavyweight rifles chambering the 'full power' 7.62x51 round. This is a British L1A1 (modified FN FAL), fitted with a Pilkington PE Snipe electro-optical sight.

taking, as it became apparent that none of the other experimental assault rifles came close to matching the Kalashnikov prototype.

The 1947 No. 1 prototype differed considerably from the 1946 pattern, gaining the familiar closed-top breech cover with an ejection port on the right side. A combination safety/selector lever was fitted to the right side of the receiver, above the trigger, possibly inspired by a similar feature on the abortive Sudaev avtomat. The butt was inserted in the receiver; the barrel had been shortened; and the fore-end was noticeably longer than the barrel-top guard. The design of the trigger system had also been refined.

After about five prototypes had been made in Kovrov, a pre-production batch of rifles was made in Tula in 1948. Field trials confirmed that the design was simple, handled well and operated satisfactorily under virtually any conditions. In 1949, therefore, the gun was adopted by the Soviet Army as the '7.62mm Kalashnikov assault rifle (AK)'.

However, the AK presented a considerable manufacturing challenge that the contemporary Russian small arms industry seemed incapable of meeting.

Among the different approaches to assault rifle design are those taken by the French, with the extraordinary 5.56mm FA MAS — colloquially known as Le Clairon (The Bugle) — and by the Austrians, with the largely synthetic Steyr AUG (shown). Although each has its champions, it is difficult to imagine that either will challenge the long-term success and durability of the Kalashnikov.

Consequently, the Simonov carbine, which had been ordered into immediate mass production as the SKS-45, was also produced in quantity while experiments with the Kalashnikov rifle continued. The first series-made AK-47 did not reach the Soviet Army much before the end of 1949.

Fifty years later, the AK is still highly regarded as a military weapon, serving in many guises throughout the world. Production of Kalashnikov and lineal derivatives continues, having exceeded seventy million.

It has been suggested that the Kalashnikov is to be replaced in Russian service by the AN-94 (Nikonov) assault rifle when funds permit, and it will be interesting to see if the AN — touted as the 'New Kalashnikov' — is really good enough to supersede the AK-74. Some reports indicate that the Nikonov sacrifices simplicity, one of the greatest strengths of the Kalashnikov, in pursuit of improvements that may turn out to be more theoretical than practical.

The Kalashnikov design bureau continues to promote its well-proven weapons and has developed a variant of the AK-74M — entered in the Russian Abakan trials of the early 1990s — as the basis for the 'Hundred Series' guns; which were announced in 1995. The assault rifle has also been adapted to produce the popular range of Sayga semi-automatic shotguns.

THE GUNS

THE GUNS

The descriptions of the many variants of the AK have been organised on the basis of calibre, which generally is the easiest of all the various identifying characteristics to interpret. The original Kalashnikov and the modernised version (AKM) chambered the 7.62x39mm M43 cartridge; details of these guns begin on page 60.

The many 7.62mm guns are separated by the length of the cartridge case (eg 7.62x39, 7.62x51 and 7.62x54R). The Sayga shotguns (.410, .615 and .729) follow on from the Kalashnikov-pattern assault rifles and the associated light machine-guns.

Operating principles

Kalashnikovs are surprisingly simple in operation, which is one of the keys to their exceptional success. The accompanying drawing (taken from a Soviet military manual) illustrates the parts of the mechanism at rest, with an empty box magazine.

Before firing the gun, it is necessary to replace the magazine, which can be done simply by pressing the catch (8). Detaching a magazine is easy, but replacing it requires the rear upper edge to be tipped backward and pushed upward simultaneously — a movement that soon becomes second nature.

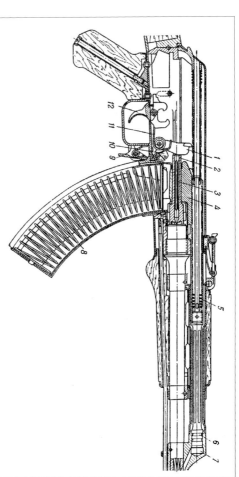

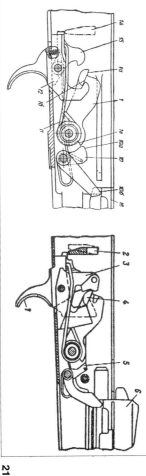

Operating mechanism of the Kalashnikov (above); trigger mechanism of the AK (below left); trigger mechanism of the AKM (below right).

To load the gun, set the selector lever on the right-hand side of the receiver to single-shot or automatic fire, then pull back the charging handle protruding from the right side of the gun (above the magazine) as far as it will go. As the handle is released to run forward, the topmost cartridge in the magazine is pushed into the chamber.

Pressing the trigger (12) releases the hammer (1), provided that the bolt is fully forward and properly locked. Assuming that the mechanism has been set for single-shot fire, the hammer flies forward to strike the firing pin (3) running through the bolt. The pin hits the primer of the chambered cartridge, and the gun fires.

The bullet travels along the barrel towards the muzzle, being pushed by propellant gas at extremely high pressure. At about two-thirds of the distance to the muzzle, the bullet passes the gas port, allowing a small amount of the propellant gas to divert into the gas cylinder (7), where it strikes the head of the piston (6) before being vented to the atmosphere. The pressure applied to the piston is transferred to the bolt carrier (4) and, as the piston and carrier move back together, a cam-way beneath the front of the carrier acts in concert with a lug on the bolt to revolve the locking lugs out of their seats in the receiver. The bolt, carrier and piston rod all move back, riding over the hammer and forcing it down until it is

held by the trigger unit. A disconnector ensures that the trigger cannot release the hammer until the firer releases his pressure on the trigger lever, then presses it again.

The spent cartridge case is extracted from the chamber as the breech opens, then ejected up to the right and out of the gun. The backward movement of the bolt, carrier and piston rod is arrested by the return spring (5), which lies in the hollow piston-rod extension. As the parts move forward again, the front lower edge of the bolt catches on the base of the topmost cartridge in the magazine and propels it into the chamber. The bolt stops against the breech face, but the carrier and the piston rod continue to move forward until the carrier cam-way revolves the bolt car-lug to ensure that the locking lugs on the bolt head securely engage their seats in the receiver. A safety interlock prevents the hammer from being released until the bolt is properly closed.

Undoubtedly, the Kalashnikov firing mechanism (below left, page 21) was inspired by that of the Garand. It comprises: the trigger (12) and the disconnector (15), sharing the same axis pin; the hammer (1); a two-armed automatic sear (10), operated by a shoulder on the bolt carrier; and two piano-wire springs, one driving the hammer and the trigger, the other for the sear and disconnector.

If fully automatic fire has been chosen, the selector prevents the disconnector from acting on the trigger, the hammer being released by the automatic sear as long as the bolt is securely locked. Firing will only stop when pressure on the trigger lever is released.

In the AKM, the Soviets revised the trigger system (below right, page 21) so completely that none of its parts is shared with the original AK. The principal addition is a hammer delay (4) with an articulated nose. This reduces the cyclic rate by forcing the hammer to overcome additional resistance before it can be released by the automatic sear. Although adding only a few milliseconds to each shot, this is enough to reduce the cyclic rate by about twenty per cent. The AKM is substantially lighter than its predecessors and would have been more difficult to control when firing automatically if the cyclic rate had not been changed.

Having the same size, weight and cyclic rate as the Shpagin submachine-gun (PPSh), the Kalashnikov assault rifle offers double the effective range, better ballistics and far greater accuracy than any gun chambered for a pistol cartridge. Although widely criticised for its clumsiness, low muzzle velocity and a poor-performing cartridge — often by people who have never fired it — the Kalashnikov is simple, solid, reliable and surprisingly effective when firing automatically.

5.45x39

Tests undertaken with M16 rifles captured in Vietnam in the 1960s showed that the ballistics of the 7.62x39 M43 cartridge were much poorer than its 5.56x45 American rival. Work continued in the USSR until 1973, when a 5.45x39 cartridge, based on the 7.62mm case, was perfected by a group headed by Viktor Sabelnikov.

Loaded 5.45x39 ball rounds measured about 57mm overall. They contained a 3.4gm bullet, 25.5mm long, comprising a two-piece core within the jacket. A small hollow tip deformed upon striking the target, allowing the bullet to topple end-over-end, increasing its destructive potential.

Good sectional density gave a flat trajectory and allowed the small-calibre projectile to pierce 5mm of steel sheet at a distance of 350m. One benefit of the comparatively low recoil impulse of the light, fast-moving bullet was that automatic fire was surprisingly easy to control.

The basic patterns

5.45mm AK-74 assault rifle
Avtomat Kalashnikova obr. 74 (AK-74)
Made by the ordnance factories in Tula and Izhevsk.

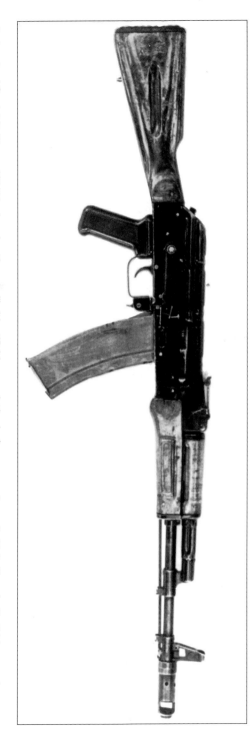

A standard Soviet 5.45mm AK-74, showing the readily identifiable muzzle-brake/compensator unit. Note the plastic thirty-round magazine and the groove in the butt.

Taken in January 1981, this posed photograph — captioned, 'No frost can stop training in the Soviet Armed Forces' — shows Lieutenant Aleksandr Bakharev giving orders to his platoon of mobile rocket launchers. His orderly carries an AK-74, identifiable by its muzzle-brake/compensator.

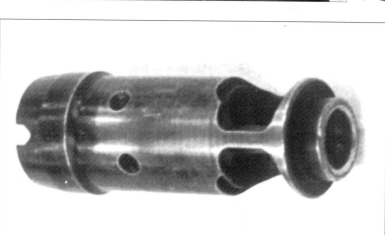

The ports in the original AK-74 muzzle-brake/compensator are shaped to deflect some of the gas forward. This minimises sideways blast, which can discomfit anyone alongside the firer.

Specification

Data for current (1998) Izhmash AO production

Calibre 5.45mm (.215)
Cartridge 5.45x39 Soviet M74 rimless
Operation Gas-operated, selective fire
Length 943mm (1089mm with bayonet fixed)
Weight 3.1kg, without magazine
Barrel 415mm, four grooves, right-hand twist
Magazine Detachable thirty-round box
Rate of fire 600±30rds/min
Selector markings 'AB' above 'ОД'
Muzzle velocity 900m/sec
Bayonet Detachable knife pattern

A small-calibre replacement for the 7.62mm AKM (which provided more than half of its parts), the AK-74 has a large cylindrical muzzle-brake/compensator that eventually gained two ports, the larger of the two on the left. The ports are angled to counteract the tendency of the gun to climb to the right when firing automatically, and also facilitate cleaning. Originally, the butt, beavertail fore-end and hand guard were laminated wood or resin-impregnated wood fibres. Longitudinal grooves cut into both sides of the butt allowed the calibre to be identified by touch. The cleaning rod could only be removed after the muzzle-brake had been detached, while two lugs beneath the barrel accepted a bayonet, even if the brake had been removed. A 1000m back sight was standard.

A typical rifle will display the standard Cyrillic selector markings, with the date and serial number (eg '1977 994048') behind the Izhevsk arrow-in-triangle mark on the left side of the receiver. The number '048' will be applied to the front left-hand end of the bolt cover. Additional marks include '33/1' on the top left side of the plastic pistol grip, with the arrow-in-triangle mark and '74' moulded into the sides of the red-brown plastic magazine. A slot will have been cut through the butt, and the rear swivel will lie on the under-edge.

Guns in current production (1999) have fixed black or grey-black plastic butts with a groove in each side and a swivel on the under-edge. The plastic pistol grip incorporates a moulded chequered panel, while the fore-end, ribbed horizontally and vertically for strength, has a prominent rounded ledge on each side to improve hand grip. The muzzle-brake/compensator currently has a single large aperture cut through the front of its body, ahead of three small radial ports.

A bracket for the 1-PN29 telescope sight can be attached to the left side of the receiver. A single-shot GP-25 grenade launcher and an associated back sight may be fitted to the fore-end and gas tube respectively. The standard accessories consist of a webbing

sling, an oiler and a small cylindrical container for the cleaning tools.

Similar guns

AK-74N (AK-74H in Cyrillic) This is the original army designation for an AK-74 with a bracket on the lower left side of the receiver for mounting a passive infra-red night sight weighing about 2.2kg. This fitting is now standard, and the 'N' suffix has been abandoned.

AKS-74 (*Avtomat Kalashnikova skladyvayushimsya prikladom obr. 74 AKC-74* in Cyrillic). Designed for use by commando and other specialised units, this model of Kalashnikov features a triangular skeleton butt, instead of the single-strut AKMS pattern, to increase rigidity. The AKS-74 butt folds forward against the left-hand side of the receiver, reducing the weapon's overall length to about 700mm.

A group of Soviet Marines parades on May Day with their AKS-74 rifles.

The current (1999) version of the AK-74.

AKS-74N (*AKC-74H* in Cyrillic) AKS-74 rifles can often mount passive infra-red sights weighing 2.2kg, which attach to a bracket on the lower left side of the receiver. These guns were originally given an 'N' suffix, but the sight rail eventually became a standard fitting, removing the need for this distinction.

Derivatives
Armenia

5.45mm Model 3 assault rifle
Manufacturer unknown.

Specification
Calibre 5.45mm (.215)
Cartridge 5.45x39 Soviet M74 rimless
Operation Gas-operated, selective fire
Length 700mm
Weight 4.0kg, with empty magazine
Barrel 415mm, four grooves,
right-hand twist

Magazine Detachable thirty-round box.
Rate of fire 600±30rds/min
Selector markings 'AB' above 'ОД'
Muzzle velocity 710m/sec
Bayonet None

Unveiled in the autumn of 1996, this is a simple adaptation of the AK-74, with the box magazine behind the pistol grip and a shoulder plate attached to the back of the black-phosphated receiver. The pistol grip and the short, vertically-ribbed fore-end are made of a dark green plastic material. Pinning of the bolt carrier to the piston rod, the only significant difference in the operating system, is purely a manufacturing expedient.

The 'bullpup' layout of this rifle substantially reduces overall length, but requires the sights to be carried on riser blocks; optional optical or electro-optical sights can be mounted in a bracket attached to the left-hand side of the receiver.

Bulgaria

5.45mm AKK-74 assault rifle
Made by the state firearms factory, Arsenal.

Specification
Data generally as Soviet AK-74 (q.v.).

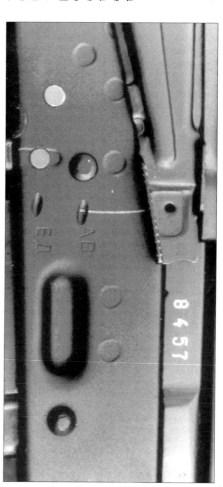

Selector and manufacturer's markings on a Bulgarian 5.45mm AKK-74. Note the ribbed plastic fore-end.

The Bulgarian-made AKM was replaced in the mid-1980s by a straightforward copy of the Soviet AK-74. Apparently, this has only been made in small quantities and can be identified largely by its marks. Like its Soviet and Russian equivalents, the selector of the Bulgarian AKK-74 is marked in Cyrillic, but the markings read 'ЕД' instead of 'ОД'

Similar guns
The AKK-74 has also been offered with a folding butt (AKKS-74), and with a bracket on the left side of the receiver for an optical or electro-optical sight (AKK-74N).

People's Republic of China

5.45mm Type 81 assault rifle
Made by China North Industries Corporation (Norinco).

Specification
Generally as Soviet AK-74 (q.v.).

Introduced in 1982, this adaptation of the AKM has a fixed butt. It is similar to the current 7.62mm Type 56 (q.v.) and has also been offered in 5.56x45 chambering. A derivative of the Type 63 rifle has been offered as the Type 81 semi-automatic carbine, but the Kalashnikov is regarded as the Type 81 submachine-gun — a distinction carried over from earlier Type 56 guns.

Similar guns
Type 81-1 This is identical to the original Type 81, but has a folding 'wrist-insert' butt of the type associated with the 7.62mm Type 56-2 (q.v.).

German Democratic Republic

5.45mm MPi-K 74 assault rifle
Maschinenpistole Kalashnikow Modell 74 (MPi-K 74)
Made by VEB Fahrzeug- und Waffenfabrik 'Ernst Thälmann', Suhl.

Specification
Generally as Soviet AK-74, except selector markings ('D' and 'E').

A copy of the AK-74 was introduced in about 1983, but production of this gun was stopped by the reunification of Germany and does not appear to have been extensive.

Hungary

5.45mm NGM-81 assault rifle
Made by Fegyver é Gázkészülékgyár (FÉG), Budapest.

Specification
Generally as Soviet AK-74, except selector markings ('1' and '∞').

Production of a Hungarian variant of the AK-74 began in the early 1980s, although only a few guns were made. Instead, effort was concentrated on a good-quality 5.56x45 variant (q.v.) intended for export. Made only with a fixed butt, a flared-base wooden pistol grip and a wooden fore-end, the NGM-81 also has a chromium-lined bore.

Poland

5.45mm KA-88 Tantal assault rifle
Karabinek automatyczny wz.88 (Kbk. wz. 88)
Made by Zaklady Metalowe Lucznik, Radom.

Specification
Data from manufacturer's literature
Calibre 5.45mm (.215)
Cartridge 5.45x39 Soviet M74 rimless
Operation Gas-operated, selective fire
Length 943mm (butt extended), 742mm (butt folded)
Weight 3.4kg, without magazine
Barrel 423mm, four grooves, right-hand twist
Magazine Detachable thirty-round box
Rate of fire 700±40rds/min
Selector markings 'C' above 'P'
Muzzle velocity 880m/sec
Bayonet Optional

Designed by a team led by Bogdan Szpaderski, the KA-88 is a modified version of the AK-74S with an additional three-round burst-firing capability. The gas-port assembly has been altered, while the selector-lever settings are repeated on the left side of the receiver, above the trigger.

The German-style folding stock has been retained, while the gun can be supplied with a wood-laminate or ribbed plastic fore-end, the injection-moulded

The Polish KA-89 Onyx submachine-gun.

pistol grip being orange-brown or dark grey-black respectively. Accessories include a bipod, Tritium night sights, telescope and electro-optical sights, collimating sights and laser range designators. The muzzle-brake doubles as a grenade launcher, although the single-shot, 40mm, bolt-action Pallad grenade launcher (40x46 Kbk-g wz. 74) together with appropriate sights can also be fitted.

Made by Zakłady Metalowe Łucznik, Radom.

5.45mm KA-89 Onyx
submachine-gun
Karabinek automatyczny wz. 89
(Kbk. wz. 89)

Specification
Data from manufacturer's literature
Calibre 5.45mm (.215)
Cartridge 5.45x39 Soviet M74 rimless
Operation Gas-operated, selective fire
Length 720mm (butt extended), 519mm (butt folded)
Weight 2.9kg, without magazine
Barrel 207mm, four grooves, right-hand twist
Magazine Detachable thirty-round box
Rate of fire 700±40rds/min
Selector markings 'C' above 'P'
Muzzle velocity 700m/sec
Bayonet None

This is a short-barrel version of the KA-88, retaining the ability to fire three-round bursts. Sales have been poor, however, and current work is being concentrated on the 5.56mm patterns in a bid to attract export sales.

Romania

5.45mm Al-74 assault rifle
Made by the state firearms factory, Cugir.

Specification
Calibre 5.45mm (.215)
Cartridge 5.45x39 Soviet M74 rimless
Operation Gas-operated, selective fire
Length 940mm
Weight 3.4kg, without magazine
Barrel 415mm, four grooves, right-hand twist
Magazine Detachable thirty-round box
Rate of fire 700±50rds/min
Selector markings 'S' above 'FA' above 'FF'
Muzzle velocity 880m/sec
Bayonet Optional

This is an indigenous version of the Soviet AK-74. Its synthetic hand guard and fore-end extend forward as far as the gas-port block, while an auxiliary pistol grip is fitted beneath the fore-end. The butt may be wood or a fixed steel skeleton. Selector markings on export guns — some of which apparently are

being offered in 5.56mm calibre — may read '1', '2' above '3'.

Russia

5.45mm AK-105 submachine-gun
Avtomat Kalashnikova obr. 105 (AK-l05)
Made by Izhmash AO, Izhevsk, Udmurt Republic.

Specification
Calibre 5.45mm (.215)
Cartridge 5.45x39 Soviet M74 rimless

Operation Gas-operated, selective fire
Length 824mm (butt extended), 586mm (butt folded)
Weight 3.0kg, without magazine
Barrel 295mm, four grooves, right-hand twist
Magazine Detachable thirty-round box
Rate of fire 650±50 rds/min
Selector markings 'AB' above 'ОД'
Muzzle velocity 840m/sec
Bayonet Detachable knife pattern

Apart from its chambering, this short-barrelled gun is basically the same as the

The Russian AK-105, currently being made by Izhmash.

7.62x39 AK-104 (q.v.), having the same plastic furniture and 500m back sight. A ready means of recognition is provided by the body of the plastic magazine, which is much straighter than the 7.62mm pattern.

5.45mm AKT assault rifle
Avtomat Kalashnikova tulskiy obratzsa (AKT)
Made by the Tula ordnance factory.

Specification
Unknown

This was an unsuccessful derivative of the Kalashnikov developed, during 1988–91, for inclusion in the Abakan assault rifle

trials. Modifications included a longer gas tube that extended virtually to the muzzle, where the take-off port was amalgamated with the front-sight block. This reduced the violence of the action, allowing the operating parts to be lightened. The breech cover was simplified, and the butt sides were recessed to save further weight, but only a few prototypes appear to have been manufactured.

USSR

5.45mm RPK-74 light machine-gun
Ruchnoi Pulemet Kalashnikova obr. 1974g (RPK-74)

Probably made in the Tula and Izhevsk arms factories.

Specification
Calibre 5.45mm (.215)
Cartridge 5.45x39 Soviet M74 rimless
Operation Gas-operated, selective fire
Length 1060mm
Weight 5.64kg, with bipod
Barrel 590mm, four grooves, right hand twist
Magazine Detachable box: thirty, forty or forty-five rounds
Rate of fire 850±50rds/min
Selector markings 'AB' above 'ОД'
Muzzle velocity 925m/sec

The RPK-74 is a 5.45mm derivative of the 7.62mm version.

The 5.45mm AKS-74U is the standard submachine-gun derivative of the locked-breech Russian Kalashnikov.

The introduction of the 5.45mm 'low-impulse' cartridge, together with the AK-74 assault rifle, allowed the development of new light machine-guns. Compared with the AK-74, the RPK-74 (which entered service in 1976) has a longer and heavier barrel, a bipod, a laterally-adjustable 1000m back sight and an RPD-like butt.

Similar guns

RPK-74N (*РПK-74H* in Cyrillic) Originally, guns equipped with night sights were given this designation, but the sight-rail has since been standardised.

RPKS-74 (*Ruchnoi Pulemet Kalashnikova*

skladyvayushimsya prikladom obr. 1974g — РПKC-74 in Cyrillic) This is a minor variant of the standard RPK-74 (q.v.) with a folding butt.

RPKS-74N (*РПK-74H* in Cyrillic) This gun was fitted with a night sight, although the designation has since been dropped, as all post-1990 examples have appropriate sight-rails.

5.45mm AKS-74U submachine-gun
Avtomat Kalashnikova skladyvayushmsya prikladom obr. 1974g ustarovka obratsza (AKS-74U)
Made by the Tula ordnance factory.

Specification
Calibre 5.45mm (.215)
Cartridge 5.45x39 Soviet M74 rimless
Operation Gas-operated, selective fire
Length 675mm (butt extended), 420mm (butt folded)
Weight 2.7kg, without magazine
Barrel 200mm, four grooves, right-hand twist
Magazine Detachable thirty-round box
Rate of fire 800±40rds/min
Selector markings 'AB' above 'РД'
Muzzle velocity 735m/sec
Bayonet None

Issued to commandos, communication

teams, sappers, tank drivers, rocket launcher crews and special police units, the short-barrelled AKS-74U bears the same relationship to the AK-74 as the AKM-SU (q.v.) does to the AKM. Its gas port has been moved back, the front-sight block altered, and the original 1000m back sight replaced with a frame containing pivoting notches for 100/200m (marked 'Π') and 400/500m (marked '4–5'). The front sight can be adjusted by screwing it up or down.

The design of the expansion chamber/flash hider has been altered, while the folding butt follows the open-triangle arrangement of the AKS-74. The butt can be swung to the left once the catch on the left-hand side of the fixture, behind the pistol grip, has been pressed. A small, sliding spring-latch situated at the front left-hand side of the receiver locks over the shoulder piece to hold the folded butt in position.

A typical AKS-74U will be marked '☆82 251285' on the left side of the receiver, with '285' repeated on the left side of the fore-end. Inspectors' marks take the form of small encircled Cyrillic letters; '2-1' will be moulded into the top left side of the pistol grip, with '☆' on a shield and '28' on the red-brown plastic body of the magazine.

Production of the AKS-74U ceased in Tula in 1997.

Similar guns

AKS-74UN (*AKC-74УH* in Cyrillic) This modified version of the short-barrelled weapon mounts a passive infra-red sight. Owing to the shortness of its barrel, the AKS-74UN is best used at comparatively short ranges.

AKS-74Y Distinguished by a semi-integral silencer attached to its shortened barrel, this is actually a version of the AKS-74U (see AKS-74UB). The confusion has arisen through an erroneous transliteration of the Cyrillic designation *AKC-74У*.

5.45mm AK-74M assault rifle
Avtomat Kalashnikova obr. 1974g modernizovanniya (AK-74M)
Made by the Izhevsk ordnance factory.

Specification
Calibre 5.45mm (.215)
Cartridge 5.45x39 Soviet M74 rimless
Operation Gas-operated, selective fire
Length 943mm (stock extended), 700mm (stock folded)
Weight 3.4kg, with loaded thirty-round magazine
Barrel 415mm, four grooves, right-hand twist
Magazine Detachable box: thirty or forty-five rounds
Rate of fire 900±50rds/min
Selector markings 'AB' above 'ОД'
Muzzle velocity 900m/sec
Bayonet Detachable knife pattern

Service experience suggested that a single, universal-issue weapon could be substituted for the fixed- and folding-butt versions of the AK-74. Thus, the butt of the AK-74M retains the conventional shape, but can be swung to the left to lie alongside the receiver; rigidity in the extended position — the weakness of many stocks of this type — is improved by a special cam-lock catch mechanism.

The fore-end and hand guard are of wear-resistant plastic, reducing the likelihood of splitting if the gun is ever used as a club in hand-to-hand combat. A weakness in the muzzle-brake/compensator attachment was corrected by lengthening the threaded attachment, while alterations to the ports cut into the body of the compensator facilitate cleaning. An amended receiver design improves rigidity, and there is a simpler means of attaching an under-barrel grenade launcher.

A rail on the left side of the receiver will accept the 1-PN29 optical, 1-PN58-2 intensifier sight, infra-red or 1-PN51 passive addressing one of the major criticisms of the original AK-74.

Developed in 1987, the AK-74M was entered in the Abakan competition to determine the future assault rifle requirements of the Russian Army. However, it does not seem to have been made in large numbers pending service evaluation of the AN-94 (Nikonov) rifle. Work is believed to have stopped in 1992.

5.5x43

Better known as .222 Remington, this chambering was restricted to the semi-automatic 'Law Enforcement' version of the 7.62x39 Finnish m/76 rifle (q.v.). Production is assumed to have been minimal, owing to the greater success of the 5.56mm alternative.

5.56x45

Developed in the USA, the 5.56x45 cartridge has become the standard small-calibre ammunition used by military and security forces in the Western world. Consequently, many of the countries that formerly were allied to — or at least were sympathetic to — the USSR have adapted the Kalashnikov and are vying with each other in search of export sales. The Yugoslavs have always offered Zastava (ZCZ) Kalashnikovs in a selection of chamberings, including 5.56x45, 7.62x51 and 7.92x57, but now have been joined in an already crowded market-place by countries such as Romania and Bulgaria.

People's Republic of China

5.56mm Type 81 assault rifle
Made by China North Industries Corporation (Norinco).

Specification
Type 81-1 This is little more than a Type 81 with a folding 'wrist-insert' butt of the type associated with the Chinese Type 56-2 (q.v.).

5.56mm Type 86 assault rifle
Made by China North Industries Corporation (Norinco).

Specification
Calibre 5.56mm (.223)
Cartridge 5.56x45 rimless
Operation Gas-operated, selective fire
Length 723mm
Weight 3.4kg, without magazine
Barrel 415mm, four grooves, right-hand twist
Magazine Detachable thirty-round box
Rate of fire 600±30rds/min
Selector markings Unknown
Muzzle velocity 900m/sec
Bayonet Probably optional

A 'bullpup' adaptation of the Type 81 series was announced in the late 1980s, but it is doubtful that series production has ever been undertaken: few have been seen in the West. A small number of guns may have been made in 5.45mm, and perhaps even in 7.62x39, but details are lacking.

Croatia

5.56mm APS-95 assault rifle
Manufacturer unconfirmed

Specification
Calibre 5.56mm (.223)
Cartridge 5.56x45 rimless
Operation Gas-operated, selective fire
Length 980mm (butt extended), 730mm (butt folded)
Weight 3.7kg, with empty magazine
Barrel 450mm, six grooves, right-hand twist
Magazine Detachable thirty-five-round box
Rate of fire 650±50rds/min
Selector markings Unknown
Muzzle velocity 915m/sec
Bayonet Probably optinal

Announced in 1996, this assault rifle is a derivative of the Galil and, thus, a direct descendant of the Kalashnikov, having the same rotating bolt and gas-piston system. The synthetic fore-end has ventilation slots, the butt folds along the right side of the receiver, and the optical sight is combined with a carrying handle. A grenade launcher is usually fitted to the muzzle of the gun.

Similar guns

Also made with 5.45mm chambering (q.v.), this adaptation of the AK-74 was introduced about 1982. It has a fixed butt, but otherwise is similar to the current Type 56 (q.v.).

Generally as Soviet AK-74 (q.v.).

Specification

Finland

5.56mm Model 71 assault rifle
Rynnakkokivääri m/71 (RK m/71)
Made by Valmet Oy, Jyväskylä.

Specification
Calibre 5.56mm (.223)
Cartridge 5.56x45 rimless
Operation Gas-operated, selective fire
Length 945mm
Weight 4.0kg, with empty magazine
Barrel 420mm, six grooves, right-hand twist
Magazine Detachable box: twenty or thirty rounds
Rate of fire 625±50rds/min
Selector markings '•••' above '•'
Muzzle velocity 975m/sec
Bayonet Detachable knife pattern

This is a variant of the standard M71 (q.v.), chambered for the US 5.56mm cartridge instead of the 7.62mm Soviet-type M43. The primary identifying feature of this assault rifle is the magazine, which is straighter than the sharply-curved 7.62mm pattern.

5.56mm Model 76 assault rifle
Rynnakkokivääri m/62/76 (RK m/62/76)

Made by Valmet Oy and Sako-Valmet, Jyväskylä.

Specification
Calibre 5.56mm (.223)
Cartridge 5.56x45 rimless
Operation Gas-operated, selective fire
Length 950mm (butt extended), 710mm (butt folded)
Weight 3.9kg, with empty magazine
Barrel 420mm, six grooves, right-hand twist
Magazine Detachable box: fifteen or thirty rounds
Rate of fire 600±25rds/min
Selector markings '•••' above '•'
Muzzle velocity 975m/sec
Bayonet Optional

This is similar to the standard 7.62mm M62/76 PT (q.v.), offered in most — if not

A Finnish Valmet-made 5.56mm m/71 assault rifle, distinguishable from its 7.62mm equivalent by the shape of the magazine, which is much straighter.

all — of the same sub-varieties, with tubular folding, fixed plastic or fixed wood butts. Usually, these guns can be recognised by the body of the plastic magazine, which is noticeably straighter than the sharply-curved 7.62mm pattern. Standard accessories include a sling and cleaning equipment; a detachable knife bayonet, blank-firing attachment and a bolt hold-open are among the optional extras.

5.56mm Model 78 light machine-gun
Konekivääri m/78
Made by Valmet Oy, Jyväskylä.

Specification
Calibre 5.56mm (.223)
Cartridge 5.56x45 rimless
Operation Gas-operated, selective fire
Length 1060mm
Weight 5.72kg, with loaded magazine
Barrel 570mm, six grooves, right-hand twist
Magazine Detachable fifteen- or thirty-round box, or 75-round drum
Rate of fire 600±50rds/min
Selector markings '•••' above '•'
Muzzle velocity 975m/sec
Bayonet None

Basically, this is an M76 assault rifle with a heavy barrel and a bipod. The sight has been moved from the bolt cover to the receiver, above the chamber, and the front sight attached to the muzzle instead of the gas-port housing. The wooden fore-end has two ventilation slots, while a folding carrying handle is fitted ahead of the back sight, and the butt has a deep belly.

Among the optional extras are a bayonet, a blank-firing attachment and a bolt hold-open. These 5.56mm guns are often fitted with a recoil buffer.

Similar guns
M78S This is a heavy-barrel 5.56mm 'Law Enforcement' rifle, restricted to semi-automatic fire, but structurally identical to the light machine-gun.

5.56mm Model 90 assault rifle
Rynnakkokivääri m/90 PT (RK m/90 PT)
Made by Sako Oy, Jyväskylä.

Specification
Data from manufacturer's literature
Calibre 5.56mm (.223)
Cartridge 5.56x45 rimless
Operation Gas-operated, selective fire
Length 898mm (butt extended), 644mm (butt folded)
Weight 3.7kg, with empty magazine
Barrel 386mm, six grooves, right-hand twist
Magazine Detachable thirty-round box
Rate of fire 600±25rds/min
Selector markings '•••' above '•'
Muzzle velocity 950m/sec
Bayonet None

Made only in small quantities, this is a small-calibre version of the standard 7.62mm rifle with a short barrel and a different rifling profile. It shares the combination front sight/gas-port housing of the large-calibre guns, has a sling loop on the underside and retains the customary two-strut tubular butt. The aperture-type, tangent-leaf back sight is graduated for 150, 300 and 400m.

Hungary

5.56mm NGM assault rifle
Mace by Fegyver é Gázkészülékgyár (FÉG), Budapest.

Specification
Generally as 5.45mm version (q.v.)

This is a minor variant of the Hungarian-made AKM rifle, chambered for the standard US cartridge instead of the customary Soviet pattern. Readily distinguished by wooden furniture and a distinctive flared-base pistol grip, the 5.56mm NGM assault rifle also has a straighter magazine body than its 5.45mm equivalent.

Israel

5.56mm ARM (Galil) assault rifle
Made by Israeli Military Industries, Ramat ha-Sharon.

The 5.56mm Hungarian NGM-81 assault rifle has a distinctively-shaped pistol grip and a tapering flash suppressor.

Specification

Data from manufacturer's literature (1984)

Calibre 5.56mm (.223)

Cartridge 5.56x45 rimless

Operation Gas-operated, selective fire

Length 979mm (butt extended), 742mm (butt folded)

Weight 4.35kg, with empty magazine and bipod

Barrel 460mm without flash suppressor, six grooves, right-hand twist

Magazine Detachable box: twelve, twenty-five, thirty-five or fifty rounds

Rate of fire 650±50rds/min

Selector markings 'S' above 'A' above 'R'

Muzzle velocity 980m/sec

Bayonet Detachable knife pattern

The Israeli authorities decided to produce a modified Kalashnikov in the late 1960s to replace the FAL. Known as the Galil, in honour of the engineer responsible for its transformation, the Israeli rifle is similar to the Finnish M62 — indeed, the earliest guns incorporated unmarked Valmet-made receivers. The pivoting aperture sight for 300 and 500m was accompanied by a flip-up night sight with integral Tritium markers.

The ARM of 1971 has a tubular plastic pistol grip and a tubular butt that can be folded along the right side of the receiver. A radial selector is fitted on the right side, closing the ejection port in its uppermost (safe) position. The cocking handle is bent upward so that the rifle can be cocked with either hand, and an optional carrying handle is provided. Fluted wooden fore-ends appeared on the earliest guns, but later examples are nylon.

38

The 5.56mm Galil ARM, with its bayonet, rifle grenades, magazine options and loading tool. Note the carrying handle and the bipod, which distinguish the ARM from the otherwise similar AR.

Six ports in the gas-check formed in the piston rod allow a small amount of the gas tapped from the bore to flow back to 'blast clean' the moving parts. Some guns also have a night safety catch, which allows the action to be cycled without feeding a round from the magazine.

The standard ARM has a bipod, hinged to the gas block to double as a wire cutter, while a short flash suppressor/compensator is fitted to the muzzle. The 1000m back sight at the rear of the breech cover is accompanied by projecting ears and a folding 100m night sight with luminescent Tritium dots. A change in the design of the flash suppressor was made in about 1980, when the original three-slot version was superseded by an improved small-port design.

Markings on a typical gun will read 'A.R. GALIL' above '5.56x45', 'IMI' and 'ISRAEL' on the left side of the receiver. The radial selector on the right side of the receiver is duplicated by a sliding catch above the left side of the pistol grip, marked 'S', 'A' and 'R' in a horizontal line (some guns made in the 1970s have the auxiliary selector reversed with 'R', 'A' and 'S' markings). Generally, export models are marked in English. Israeli Defence Force rifles,

The standard 5.56mm Galil AR.

stamped in Hebrew, also display a sword and olive branch inside a six-pointed star on the left side of the receiver, above the pistol grip.

Guns made since the early 1980s may bear the Israeli Military Industries trademark — a sword and an olive branch superimposed on a cogwheel — and have serial numbers running vertically upward on the front left-hand side of the receiver, immediately behind the fore-end.

The twelve-round magazine holds special ammunition required for use with the grenade launcher; the thirty-five-round pattern is restricted to blanks.

5.56mm AR assault rifle
Made by Israeli Military Industries, Ramat ha-Sharon.

Specification
Generally as ARM (q.v.), except for a weight of about 3.95kg

This is a long-barrelled ARM, lacking the bipod and carrying handle. It usually has a simplified nylon fore-end, although the earliest examples were wood.

5.56mm SAR assault rifle
Made by Israeli Military Industries, Ramat ha-Sharon.

Specification
Data from manufacturer's literature (1984)
Calibre 5.56mm (.223)
Cartridge 5.56x45 rimless
Operation Gas-operated, selective fire
Length 851mm (butt extended), 614mm (butt folded)
Weight 3.75kg, with empty magazine
Barrel 332mm, six grooves, right-hand twist
Magazine Detachable box: twelve, twenty-five or fifty rounds
Rate of fire 650±50rds/min
Selector markings 'S' above 'A' above 'R'
Muzzle velocity 920m/sec
Bayonet Detachable knife pattern

41

One of the most important requirements for the Galil was that it should work effectively in desert conditions.

Essentially, the Short Assault Rifle (SAR) is a short-barrelled ARM. It lacks the bipod and carrying handle of the latter, but retains the standard skeleton butt and open sights.

5.56mm MAR carbine

Made by Israeli Military Industries, Ramat ha-Sharon.

Specification

Data from manufacturer's literature (1996)

Calibre 5.56mm (.223)

Cartridge 5.56x45 rimless

Operation Gas-operated, selective fire

Length 690mm (butt extended), 445mm (butt folded)

Weight 2.95kg, with empty magazine

Barrel 195mm, six grooves, right-hand twist

Magazine Detachable thirty-five-round box

Rate of fire 650±50rds/min

Selector markings 'S' above 'A' above 'R'

Muzzle velocity 710m/sec

Bayonet None

The Micro Assault Rifle, the shortest gun in the Galil series, is intended for special forces, vehicle crews and others who

The standard 5.56mm Galil SAR. This particular gun has a bayonet lug, which is an optional extra.

Italy

A modified version of the Israeli Galil was submitted to the Italian Army in the 1980s by Bernardelli. However, the Beretta AR 70 was preferred, and development of the Bernardelli 5.56mm SR, or SR 556, was stopped.

Poland

5.56mm KA-90 Tantal assault rifle

Also known as *Kbk. wz. 90*

Made by Zaklady Metalowe Lucznik, Radom.

would benefit from an ultra-compact design. The length of the barrel is an obvious identifying feature, but the fore-end has a prominent lip to prevent the hand from slipping forward, which would be potentially dangerous given the proximity of the muzzle to the front of the fore-end. The tubular butt swings forward to the right.

The Polish KA-90 Tantal rifle.

Specification
Data from manufacturer's literature
Calibre 5.56mm (.223)
Cartridge 5.56x45 rimless
Operation Gas-operated, selective fire
Length 943mm (butt extended), 742mm (butt folded)
Weight 3.4kg, without magazine

Barrel 423mm, four grooves, right-hand twist
Magazine Detachable thirty-round box
Rate of fire 700±25rds/min
Selector markings 'C' above 'P'
Muzzle velocity 900m/sec
Baycnet Detachable knife pattern

This variant of the 5.45mm wz. 88 has a straighter magazine with strengthening ribs, and a muzzle-brake that doubles as a grenade launcher. The German-style folding stock is retained, and an additional three-round burst-firing capability can be supplied on request. Guns may have wood-laminate or ribbed plastic fore-ends, the injection moulded pistol grips being orange-brown or dark grey-black respectively. Accessories include a bipod, Tritium night sights, telescope and electro-optical sights, collimating sights and laser range designators. The bolt-action, 40mm Pallad grenade launcher and associated sights can also be fitted to this gun.

5.56mm KA-91 Onyx submachine-gun

Also known as *Onyks* and *Kbk. wz. 91*

Made by Zaklady Metalowe Lucznik, Radom.

Specification
Data from manufacturer's literature
Calibre 5.56mm (.223)
Cartridge 5.56x45 rimless
Operation Gas-operated, selective fire
Length 720mm (butt extended), 519mm (butt folded)
Weight 2.9kg, without magazine
Barrel 207mm, four grooves, right-hand twist
Magazine Detachable thirty-round box
Rate of fire 700±25rds/min
Selector markings Various
Muzzle velocity 710m/sec
Bayonet None

Basically, the Kbk. wz. 90 is a short-barrel version of the Tantal, resembling the Soviet AKMS-U (q.v.). The short fore-end is usually made of laminated wood, although the pistol grip and barrel guard are plastic. A short cylindrical muzzle-brake is fitted with a conical flash suppressor. The short-range back sight (100/200/400m) is mounted on an extension of the sight block that is carried back above the ejection port; optional sights include a laser designator and a 'Red Dot' collimator unit. The German-style mono-strut stock can be folded forward along the right-hand side of the receiver.

5.56mm KA-96 Beryl assault rifle

Made by Zaklady Metalowe Lucznik, Radom.

Specification
Data from manufacturer's literature
Calibre 5.56mm (.223)
Cartridge 5.56x45 rimless
Operation Gas-operated, selective fire
Length 943mm (butt extended), 742mm (butt folded)
Weight 3.535kg, with empty magazine
Barrel 457mm, six grooves, right-hand twist
Magazine Detachable thirty-round box
Rate of fire 700±25rds/min
Selector markings Various
Muzzle velocity 920m/sec
Bayonet Detachable knife pattern

A modernised version of the KA-90, the Beryl assault rifle can be distinguished by the design of its folding butt — it is much more like the FNC skeleton than the previous German mono-strut favoured in Poland — and by the plastic fore-end with diagonal ribbing. Undoubtedly, the butt is more rigid than its predecessor, largely due to a longer pivot-bolt, and is less likely to be bent in combat. It also has a rubber butt pad, which improves comfort and prevents it from slipping from the firer's shoulder.

The selector gives a choice of single shots, three-shot bursts or fully automatic fire. The muzzle-brake/compensator has been reduced to a small cylinder attached to the barrel, while a 40mm Pallad grenade launcher and its sights can be attached beneath the fore-end. A special rail can be fitted above the bolt cover to accept the PCS-6 passive infra-red night sight, CK-3 collimator sight, LKA-4 optical sight or CWL-1 laser designator.

Standard accessories comprise four thirty-round magazines, a magazine pouch, four fifteen-round magazine loading chargers, a detachable bipod, a two-part rod and associated cleaning equipment, and a blank-firing attachment. A wire-cutter bayonet and a twenty-round magazine are available to order.

5.56mm KbkA-96 Mini-Beryl submachine-gun

Made by Zaklady Metalowe Lucznik, Radom.

The Polish KA-96 Beryl rifle.

The Polish KbkA-96 Mini-Beryl submachine-gun.

Specification

Data from manufacturer's literature

Calibre 5.56mm (.223)

Cartridge 5.56x45 rimless

Operation Gas-operated, selective fire

Length 730mm (butt extended), 525mm (butt folded)

Weight 3.185kg, with empty thirty-round magazine

Barrel 235mm, six grooves, right-hand twist

Magazine Detachable box: twenty or thirty rounds

Rate of fire 700±25rds/min

Selector markings Various

Muzzle velocity 770m/sec

Bayonet None.

Essentially a short version of the standard Beryl assault rifle, this gun is distinguished by its short barrel — so short, in fact, that the muzzle-brake/flash suppressor abuts the front sight, which is combined with the gas-port housing. The fore-end is of grey-black plastic, ribbed diagonally, and the skeleton-type folding butt is retained. Optical and electro-optical sights can be used with this submachine-gun in conjunction with the optional receiver-top rail.

Romania

5.56mm AI-74 assault rifle

Made by the state firearms factory, Cugir.

Specification
Generally as 5.45mm version (q.v.), except for chambering

Derived from the Soviet AK-74, the AI-74 has a hand guard that runs as far as the gas-port block, and a similarly elongated fore-end with an auxiliary pistol grip. The butt may be wood or a fixed steel skeleton type.

Selector markings 'AB' above 'ОД'
Muzzle velocity 910m/sec
Bayonet Detachable knife pattern

Russia

5.56mm AK-101 assault rifle

Avtomat Kalashnikova, obr. 101 (AK-101)
Made by Izhmash AO, Izhevsk, Udmurt Republic.

Specification
Calibre 5.56mm (.223)
Cartridge 5.56x45 rimless
Operation Gas-operated, selective fire
Length 943mm (butt extended), 700mm (butt folded)
Weight 3.4kg, without magazine
Barrel 415mm, four grooves, right-hand twist
Magazine Detachable thirty-round box
Rate of fire 600±25rds/min

The AK-101 is a minor variation of the current Izhmash-made AK-74 (q.v.). The major difference lies in the chambering, which accepts the US 5.56mm round instead of the Soviet 5.45mm pattern. Unlike most other 5.56mm-calibre rifles, which have six grooves, the AK-101 retains four-groove rifling to harmonise production with the 5.45 and 7.62mm Kalashnikovs. It is not yet known whether this has any effect on accuracy.

The most obvious external characteristic is the shape of the plastic magazine body, which is much straighter than the Russian 5.45 and 7.62mm types. The AK-101 is supplied with the standard accessories: a webbing sling, an oiler and a selection of cleaning tools in a small cylindrical container.

5.56mm AK-102 submachine-gun

Avtomat Kalashnikova obr. 102 (AK-102)
Made by Izhmash AO, Izhevsk, Udmurt Republic.

The Russian AK-101, made by Izhmash.

The Russian AK-102, a short-barrelled gun currently offered by Izhmash.

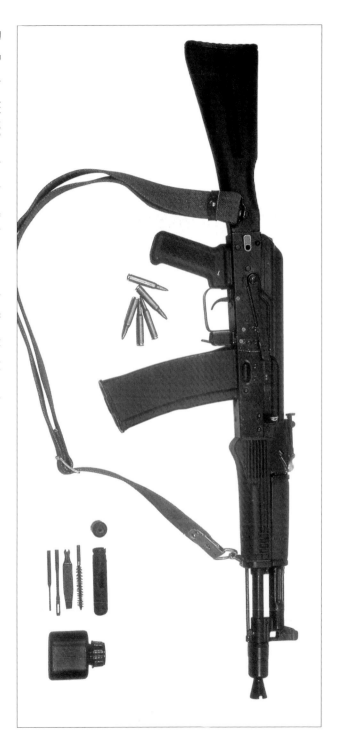

Specification

Calibre 5.56mm (.223)

Cartridge 5.56x45 rimless

Operation Gas-operated, selective fire

Length 824mm (butt extended), 586mm (butt folded)

Weight 3.0kg, without magazine

Barrel 295mm, four grooves, right-hand twist

Magazine Detachable thirty-round box

Rate of fire 600±25rds/min

Selector markings 'AB' above 'OД'

Muzzle velocity 850m/sec

Bayonet None

This is simply a short-barrelled version of the AK-101, sharing the constructional details of the AK-74 (q.v.). These include a short muzzle-brake/compensator that abuts the front sight/gas-port assembly, a 500m back sight and a rail for mounting optical or electro-optical sights on the left-hand side of the receiver. Its most obvious identifying feature is the magazine, which is made with a much straighter body than the 5.45 and 7.62mm equivalents.

Republic of South Africa

After protracted trials, the South African government adopted the 5.56mm Rifle 4 during 1982, using it to replace the 7.62mm R1 (FAL), as well as some G3 weapons that had been seized through operations in the Angolan borderland. The new rifle is a development of the Israeli Galil (q.v.).

5.56mm Rifle Model 4

Also known as R-4
Made by Lyttleton Engineering Works Pty (now Vektor), Pretoria.

Specification

Calibre 5.56mm (.223)
Cartridge 5.56x45 rimless
Operation Gas-operated, selective fire
Length 1005mm (butt extended), 740mm (butt folded)
Weight 4.3kg, without magazine or bipod
Barrel 460mm, six grooves, right-hand twist
Magazine Detachable thirty-five-round box
Rate of fire 675±75rds/min
Selector markings 'S' above 'A' above 'R'
Muzzle velocity 980m/sec
Bayonet None

The fore-end and pistol grip of the R-4 are made from fibreglass-reinforced plastic. Similarly, the butt, which is steel on the Israeli gun, is synthetic to reduce the effects of the hot South African climate. The butt has also been extended and has a much longer fillet between the two struts. Changes have been made to the optional detachable bipod, gas system, receiver and rifling to suit SADF requirements.

5.56mm Rifle Model 5

Also known as R-5
Made by Lyttleton Engineering Works Pty (now Vektor), Pretoria.

The South African 5.56mm R-4 rifle, a close copy of the Israeli Galil. Note the design of the butt, which is longer than its prototype

The South African R-4 with butt folded forward.

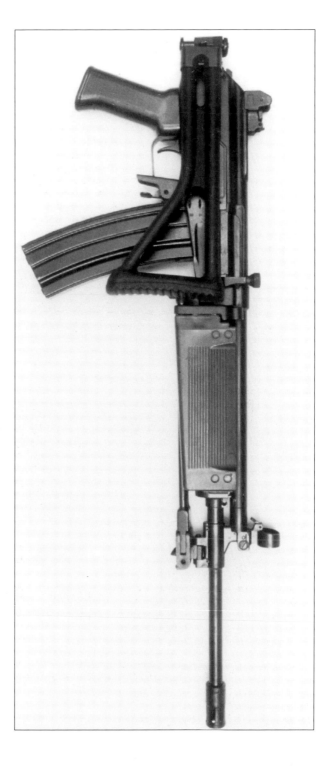

Specification

Calibre 5.56mm (.223)

Cartridge 5.56x45 rimless

Operation Gas-operated, selective fire

Length 880mm (butt extended), 615mm (butt folded)

Weight 3.7kg, without magazine

Barrel 332mm, six grooves, right-hand twist

Magazine Detachable thirty-round box

Rate of fire 675±75rds/min

Selector markings 'S' above 'A' above 'R'

Muzzle velocity 920m/sec

Bayonet None

Adopted in 1987 for the South African air force and marines, the R-5 is intended to supplement and eventually replace the R-4. In many respects, it resembles its predecessor except for the lack of a bipod and a 128mm reduction in barrel length. The gas tube, piston and hand guard are all significantly shorter than the equivalent items on the R-4 type, while the flash suppressor cannot be adapted to allow for grenade launching.

(Left) South African troops operating in the RSA/Angola borderland, with R-4 rifles and a heavy mortar.

(Right) Pictured in December 1979, these Rhodesian (Zimbabwean) guerrillas display the customary motley collection of Kalashnikovs, in this case a mixture of AKM and AKMS.

The South African 5.56mm R-5, a short-barrel version of the R-4. The gun lacks its magazine.

5.56mm Rifle Model 6

Also known as R-6 Compact
Made by Vektor Engineering, Pretoria.

Specification

Calibre 5.56mm (.223)
Cartridge 5.56x45 rimless
Operation Gas-operated, selective fire
Length 805mm (butt extended), 565mm (butt folded)
Weight 3.675kg, without magazine
Barrel 280mm, six grooves, right-hand twist
Magazine Detachable thirty-round box
Rate of fire 585±35rds/min
Selector markings 'S' above 'A' above 'R'
Muzzle velocity 825m/sec
Bayonet None

The R-6 was introduced in the early 1990s for vehicle crews, paratroops and similar units. Basically, it is an R-5 with the barrel shortened until the flash hider virtually abuts the combination front sight and gas-port housing block.

Sweden

5.56mm FFV 890C assault rifle

Made by Forenade Fabriksverken AB, Eskilstuna.

The FFV 890C is a compact version of the Galil AR, made for trials in Sweden during the 1980s. Note the design of the fore-end and the extra depth of the trigger/trigger guard to accommodate an arctic mitten. FFV's involvement with the Kalashnikov was short-lived, as the military trials favoured the FN FNC.

Specification
Calibre 5.56mm (.223)
Cartridge 5.56x45 rimless
Operation Gas-operated, selective fire
Length 860mm (butt extended), 625mm (butt folded)
Weight 3.5kg, without magazine
Barrel 340mm, six grooves, right-hand twist

Magazine Detachable thirty-five-round box
Rate of fire 650±50rds/min
Selector markings Unknown
Muzzle velocity 860m/sec
Bayonet None.

FFV entered this variant of the Galil in the Swedish Army's assault rifle trials that were held during the 1980s. The gun's rifling was changed to suit the FN SS109 bullet, a ventilated synthetic hand guard was fitted, the gas system was modified, the sights were altered, and the trigger guard was enlarged to provide room for the finger of an arctic mitten. However, the army preferred the FNC, which ultimately was adopted as the Ak-5, and the FFV version of the Kalashnikov was abandoned.

Yugoslavia/Serbia

5.56mm Model 80 assault rifle

Automatska puška vz. 80
Made by Zavodi Crvena
Zastava, Kragujevač.

Specification

Data from manufacturer's literature
Calibre 5.56mm (.223)
Cartridge 5.56x45 rimless
Operation Gas-operated, selective fire
Length 985mm
Weight 4.03kg, with empty magazine

Barrel 460mm, six grooves, right-hand twist
Magazine Detachable thirty-round box
Rate of fire 600±25rds/min
Selector markings 'U' above 'R' above 'J'
Muzzle velocity 915m/sec
Bayonet Detachable knife pattern

This rifle, introduced in the early 1980s, is an adaptation of the basic Kalashnikov design for the 5.56mm cartridge. It has a modified gas system to improve performance, and a longitudinally-slotted compensator is used. Some guns have rifling for the US M193 bullet; others accept the Belgian SS109 type, the grooves (usually six) making a turn in 7.1in and 12in respectively. The M80 has a fixed wooden butt and a synthetic pistol grip. It can be distinguished from the 7.62x39 version by its muzzle fittings and short, straight, staggered-column magazine.

Yugoslav Kalashnikovs may be encountered with .38-calibre baton-round firing attachments and special magazines capable of holding three or nine rounds. The baton-round attachment is about 300mm long and weighs 400gm.

The 5.56mm Yugoslav M80 assault rifle, with a wooden butt and fore-end.

The Yugoslav M82 light machine-gun with fixed wooden butt.

Similar guns

M80A This variant of the Yugoslavian Model 80 assault rifle is equipped with a stamped-strip butt that folds down and forward beneath the gun's receiver, reducing the overall length to 725mm with the butt folded.

5.56mm Model 82 light machine gun

Puškomitraljez vz. 82

Made by Zavodi Crvena Zastava, Kragujevac.

Specification

Calibre 5.56mm (.223)

Cartridge 5.56x45 rimless

Operation Gas-operated, selective fire

Length 1025mm

Weight 4.5kg, without magazine

Barrel 500mm, six grooves, right-hand twist

Magazine Detachable box: thirty or forty-five rounds

Rate of fire 600±25rds/min

Selector markings 'U' above 'R' above 'J'

Muzzle velocity 960m/sec

Bayonet None

An adaptation of the basic Yugoslav Kalashnikov assault rifle, the M82 light machine-gun can be distinguished by its heavy barrel and bipod. The standard version has a fixed wooden butt and a synthetic pistol grip.

Similar guns

M82A This variant of the Model 80 assault rifle is equipped with a stamped-strip butt that folds down and forward beneath the receiver. This reduces overall length to 765mm.

57

The folding-butt M82A light machine-gun.

5.56mm Model 85 assault rifle
Automatska puška vz. 85

Made by Zavodi Crvena Zastava, Kragujevac.

Specification
Data from manufacturer's literature
Calibre 5.56mm (.223)
Cartridge 5.56x45 rimless
Operation Gas-operated, selective fire
Length 790mm (butt extended), 570mm (butt folded)
Weight 3.2kg, without magazine.
Barrel 254mm, six grooves, right-hand twist
Magazine Detachable box: twenty or thirty rounds
Rate of fire 700±50rds/min
Selector markings 'U' above 'R' above 'J'
Muzzle velocity 790m/sec
Bayonet None

Inspired by the Soviet AKS-74U, this compact weapon was introduced principally for use in armoured personnel carriers. It has a standard Yugoslavian-pattern butt, which folds down and forward beneath the receiver, and a hinged breech cover that doubles as the back sight base. The rocking-'L' sight is regulated for 100 and 500m. The cylindrical expansion chamber fitted to the muzzle is simply a copy of its Soviet counterpart.

6.17x51

This is the metric designation for the .243 Winchester round, introduced as a companion to the .308 pattern (7.62x51 NATO). Given the similarity of the .243 and .308 cases, Valmet took the opportunity of chambering a version of the Petra sporting rifle (see 7.62x51) for the smaller round.

7.62x39

The Soviet 7.62mm 'intermediate' cartridge, developed in the mid-1940s, remained as the principal cartridge chambered in the myriad versions of the Kalashnikov rifle until the advent of the 5.45x39 round in the mid-1970s. Many former Soviet satellite countries, together with the People's Republic of China, have continued to make 7.62mm guns virtually to the present day. Zaklady Metalowe Lucznik in Poland, for example, still offers newly-made AKM and AKMS rifles.

The basic patterns

7.62mm AK47 assault rifle
Avtomat Kalashnikova obr. 1947g
(AK or AK-47)
Made by the ordnance factories in Tula and Izhevsk.

Specification
Calibre 7.62mm (.30)
Cartridge 7.62x39 Soviet M43 rimless
Operation Gas-operated, selective fire
Length 870mm
Weight 4.3kg, with empty magazine
Barrel 415mm, four grooves, right-hand twist
Magazine Detachable thirty-round box
Rate of fire 775±50rds/min
Selector markings 'AB' above 'ОД'
Muzzle velocity 710m/sec
Bayonet Detachable knife pattern

The earliest guns — dating from 1948–51 — were made largely from welded components, stampings and pressed-

This 1951-vintage, Tula-made AK represents the earliest stage of Soviet production, with the original pressed-steel receiver. Note the design of the pistol grip and the position of the sling swivel, which has been moved to the side of the butt from the under-edge.

metal parts. They had wooden butts and fore-ends, the pistol grips originally being laminated wood, but subsequently often replaced by plastic versions. The steel butt plate incorporated a hinged trap for the cleaning equipment, an appropriate rod being carried beneath the barrel. The back sight was an 800m type.

It has often been suggested that the earliest Kalashnikov rifles lacked a bayonet lug, inferring that the Russians had developed the ungainly knife pattern only after some guns had been made. This is now known to be untrue, since even some of the prototype guns accepted bayonets, although the first production versions did not.

Tolerably made, if not particularly well finished, a typical early rifle will have a coarsely chequered plastic pistol grip with a concentric-circle design within a lozenge-shaped escutcheon. A cleaning rod will be carried beneath the barrel, but there will be no bayonet lug.

The selector will have been crudely marked 'AB' and 'ОД' with an electric pencil, possibly at some time after manufacture, as some guns have been reported with 'ABT' instead of 'AB'. The magazine will have a plain body (subsequently, this type was replaced by a sturdier ribbed pattern). The left side of the receiver will display the date (eg '1951 Г'), with the serial number (such as '3П4014') applied separately; a small principal inspector's or proof mark in the form of two encircled crossed hammers will indicate that it was made in Tula.

Customarily, the prefixed serial number is repeated on the right side of the fore-end, ahead of the receiver, and on the back of the bolt cover. Inspectors'

The second Soviet-pattern AK has a receiver machined from a forged-steel billet, while the butt is held in a distinctive shoe that protrudes from the back of the receiver. This gun has a simpler pistol grip than the earlier example and lacks its cleaning rod; the ribbed bolt cover is not original. Selector locators take the form of dots instead of grooves, and the magazine is the reinforced-rib type.

A typical second-pattern AKS, this gun has a butt that can be swung down below the receiver. Made in the Izhevsk factory in 1952, it also relies on dots instead of grooves to distinguish the selector positions.

marks take the form of small Cyrillic characters, frequently (although not always) applied within circles, squares and diamonds.

Problems arose with the original stamped-receiver guns, and a change was made in 1951 to sturdier receivers machined from forged steel billets. This may have resulted from a switch of production from Tula to Izhevsk, reflecting differences in manufacturing techniques.

AK-47s dating from 1953 have an extension, or 'shoe', on the receiver to accept the butt. The furniture, including the pistol grip, is of laminated wood. The

clearing rod is carried beneath the barrel, a bayonet lug is fitted, and the magazine has plain sides.

The selector marks are stamped into the receiver alongside the selector itself, while the arrow-in-triangle mark of the Izhevsk factory lies ahead of the date/serial number group in the depression milled into the left side of the receiver to save weight. The prefixed number is repeated on the right side of the fore-end, and often mismatched, on the back of the bolt cover.

The third, or perfected, model, dating from 1954, has its butt attached directly

with a tongue (which enters the receiver body) and two screws that run into the wood through a tang. Undoubtedly stronger than its predecessors, this method of construction lasted until the end of production.

Similar guns

AKS *(Avtomat Kalashnikova sklady-vayushimsya prikladom obr. 1947g — AKC or AKC-47* in Cyrillic) This was an otherwise standard Kalashnikov with a pressed-steel butt, which folded down and forward under the receiver to reduce overall length to about 700mm. Its

The left side of an AKS. Note that the rear sling swivel is anchored in the butt pivot; the front attachment consists of a fixed D-loop on the gas-port block. This gun also has luminous night sights.

compact design was particularly suitable for airborne forces, tank troops and other specialist personnel.

7.62mm AKM assault rifle

Avtomat Kalashnikova Modernizirovanniya (AKM)

Made by the ordnance factories in Tula and Izhevsk.

Specification

Calibre 7.62mm (.30)

Cartridge 7.62x39 Soviet M43 rimless

Operation Gas-operated, selective fire

Length 878mm

Weight 3.85kg, without magazine

Barrel 415mm, four grooves, right-hand twist

Magazine Detachable thirty-round box

Rate of fire 650±30rds/min

Selector markings 'AB' above 'ОД'

Muzzle velocity 710m/sec

Bayonet Detachable knife pattern

Once Soviet industry had mastered appropriate metalworking techniques, a modified Kalashnikov *avtomat* was introduced in 1959. This has a stamped-steel receiver — a sturdy U-shaped pressing — that is much lighter than the original machined forging. The bolt-lock recesses are riveted in place, while the stamped receiver cover has reinforcing ribs. The gas-piston tube of the AKM has semi-circular vents immediately behind the gas-port assembly, instead of the circular holes of the Ak, while the bolt carrier is phosphated.

The charging handle and pistol grip are made of plastic, although the butt and fore-end are usually produced from laminated wood. The very earliest magazines are formed from ribbed sheet-metal, but subsequent versions are moulded in an orange-red plastic.

A typical Soviet AKM, made in Izhevsk in 1978. Note the design of the receiver, which lacks the milled panel so characteristic of the AK, and the position of the sling swivel on the butt.

The most important internal change is the incorporation of a rate-reducer, or retarder, in the trigger system. This has often been identified in English-language publications as an additional mechanical safety, but Russian handbooks make its purpose crystal clear. By holding back the hammer after the bolt-carrier has depressed the safety sear, it reduces the rate of fire by about fifteen per cent, but needlessly complicates an essentially simple design.

The impact of the bolt carrier on the bolt as the action closes was transferred from the right side (AK) to the left side (AKM) in an attempt to stabilise the gun when firing automatically, and a new 100cm back sight was adopted. A short oblique-cut compensator — added in the early 1960s — prevents the gun from climbing to the right when being fired automatically, and improves accuracy when firing from awkward positions. Fore-ends were broadened in this era to improve grip, while an insulated bayonet handle allows electrical cables to be cut in safety.

A typical AKM, with a stamped, ribbed receiver cover, has a wooden butt and a beavertail fore-end of wood laminate. The plastic pistol grip has an integrally moulded chequered panel, while the crude, reddish-orange plastic magazine has a metal lip insert and floor plate to ensure reliable feed; it also bears a small arrow-in-triangle Izhevsk mark moulded into the body.

The gun has a 1000m back sight, standard Cyrillic selector markings, a cleaning rod, a bayonet lug, and a swivel

A typical AKMS and its markings. This particular gun was made in Izhevsk in 1975. Note that the serial number lacks prefix letters.

on the under-edge of the butt. The factory mark, date and serial number are stamped on the left-hand side of the receiver, the number being repeated on the back of the bolt cover, above the release catch.

Some guns were issued with the PBS-1 silencer, which screws directly on to the muzzle once the compensator has been removed. A bracket can be added to the left side of the receiver to accept the NSP-2 infra-red or NSPU image-intensifier sights. Some assault rifles will also be found with the single-shot GP-26 grenade launcher under the fore-end and an auxiliary sight fitted above the gas tube. A special resilient butt pad was issued with these guns to minimise the effects of recoil on the firer's shoulder.

Similar guns

AKMS (*Avtomat Kalashnikova Moderniz-irovanniya, skladyvayushimsya pri-kladom — AKMC in Cyrillic*) This is a variant of the standard gun with a metal butt, characterised by three rivets and a long flute on each side of the strut. The butt can be swung down and forward once the locking catch had been released.

A typical example will have a laminated wood beavertail fore-end and a plastic, chequer-panel pistol grip. The compensator is a small oblique-cut type, it has a 1000m back sight, the cleaning rod is carried beneath the barrel, and a bayonet lug is provided. Markings include 'AB' and 'ОД' on the receiver alongside the selector, while the Izhevsk arrow-in-triangle mark appears on the front left side of the receiver, ahead of the year and serial number.

Derivatives
Bulgaria

Initially, the Bulgarian People's Army was armed with rifles made in Poland. Later

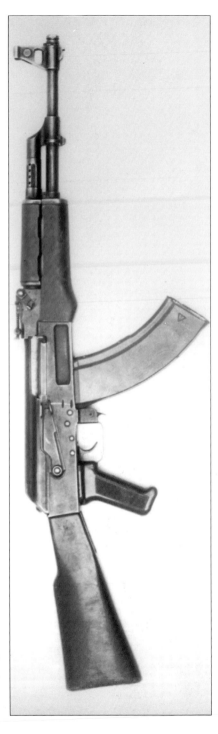

(Above and opposite) Two views of a Bulgarian AKK, fitted with a Soviet-made plastic magazine. Note the comparatively crude finish, the design of the pistol grip and the position of the swivel on the side of the receiver.

examples may have been assembled in Bulgaria from Polish-made parts, as some have been encountered with a Polish factory mark — '11' in an oval — above more typically Bulgarian identifiers; for example, '10' in a double circle ahead of '1963 BY 1645' (the date and serial number). The serial number, without the two-letter prefix, is frequently repeated on the rifle's bolt cover and the fore-end.

7.62mm AKK assault rifle

Avtomatícheskiy Karabín Kalashnikova

Made by the state firearms factory, Arsenal.

Specification

Generally similar to the Soviet AK, except crudely chequered. A cleaning rod is carried beneath the barrel.

Bulgarian production began in State Factory No. 10 in the mid-1960s. The earliest guns had machined-steel receivers, wooden fore-ends, plain wooden pistol grips, and cleaning rods beneath their barrels.

A typical example bears the figure '10' within a double circle on the left side of the receiver, ahead of the date and serial number (eg 'KT 17-4749') applied with an electric pencil; 'KT 4749' is repeated on the back of the bolt cover, with '4749' on

the fore-end, ahead of the receiver. The furniture is plastic, the pistol grip being crudely chequered. A cleaning rod is carried beneath the barrel.

One of the best identifying features is provided by the selector markings, which read 'AB' and 'EД' (instead of 'OД' on Russian/Soviet guns).

Similar guns

AKK-M1 This is a standard AKK fitted with a 40mm grenade launcher beneath the barrel and equipped with appropriate auxiliary sights.

AKKS Generally similar to the Soviet AKS, this has a solid C-section folding

This Bulgarian AKKS has an extended muzzle-cap, protecting the short screw-thread attachment for a grenade launcher or compensator.

butt actuated by a press-stud on the left side. The shoulder piece is an open U-shape and lacks visible rivets.

7.62mm AKKM assault rifle

Avtomatícheskiy Karabin Kalashnikova Modernizírovanniya
Made by the State firearms factory, Arsenal.

Specification

Generally as Soviet AKM (q.v.), except selector markings

Production switched from the AK to the AKM in the early 1970s. Bulgarian guns can be difficult to distinguish from their Soviet equivalents, the most obvious differences being their markings.

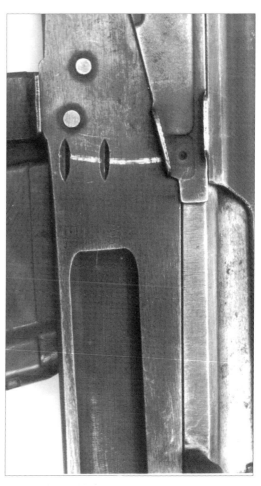

(Top Right) Bulgarian Kalashnikovs have 'AB' and 'ЕД' as selector markings, unlike their Russian equivalents, which use 'AB' and 'ОД'.

(Below right) The Bulgarian manufacturer's mark usually takes the form of a figure '10' within concentric circles. It has been suggested that the first two digits of the serial number ('18') represent the date, based on the foundation of the Socialist Republic of Bulgaria in 1948 (18 = 1965), but this seems highly unlikely.

(Above) Chinese military Kalashnikovs usually bear the mark of Factory No. 66 in a triangle. The designation '56 Type' (originally entirely in ideographs, but later partly numerical) appears ahead of an eight-digit serial number.

A Chinese Type 56 Kalashnikov. Note the position or the sling swivel and the thin gauge of its wire loop. The blade of the folding bayonet has a triangular section, although the original pattern copied the standard Soviet knife design.

People's Republic of China

7.62mm Type 56 (AK) assault rifle
Made by State Factory No. 66.

Specification

Calibre 7.62mm (.30)
Cartridge 7.62x39 Soviet M43 rimless
Operation Gas-operated, selective fire
Length 870mm
Weight 4.45kg, with empty magazine
Barrel 415mm, four grooves, right-hand twist
Magazine Detachable thirty-round box
Rate of fire 775±50rds/min

Selector markings In Chinese (military)
Muzzle velocity 720m/sec
Bayonet Integral folding pattern

The People's Republic was an early convert to the Kalashnikov, production beginning in the late 1950s. Early standard versions of the Type 56 have a plain butt, fore-end and pistol grip of wood; a receiver machined from a steel billet; and a knife-blade bayonet attached to a block beneath the muzzle.

A long, triangular-blade bayonet soon replaced the short knife pattern, although the method of attachment remained the same. In the mid-1960s, however, production was switched to a stamped-sheet AKM-type receiver without altering the basic designation (see AKM section).

A typical gun will bear the marks of Factory No. 66 — '66' in a triangle — on the left side of the receiver, ahead of the designation '56 Type' (in the form of three ideographs) and a serial number, such as '12134241'. Normally, the last five digits of the serial number are repeated on the bolt cover. Guns intended for export and commercial sale may display selectors marked 'L' and 'D', or in numerals.

This Chinese Type 56 rifle lacks its bayonet, which would have pivoted on the prominent lug beneath the muzzle, behind the front-sight block. Note that this gun has a pressed-steel AKM-type receiver and that the sling swivel is fitted beneath the butt.

71

Typical Chinese selector marks.

Similar guns

Type 56-1 This has a squared U-section butt, with an open U-shaped shoulder piece and two rivets through each side. Ideographs are usually found with the selector lever, and the factory mark — '66' in a triangle — is stamped on the side of the receiver, ahead of the designation '56-1 Type' and a serial number, such as '3013965'. The last five digits of the number are repeated on the bolt cover. The Type 56-1 assault rifle, which measures 700mm overall with its butt folded, was replaced after about 1965 by an AKM pattern (q.v.).

7.62mm Type 56 (AKM) assault rife

Made by State Factory No. 66 and an unidentified Norinco plant.

Specification

Calibre 7.62mm (.30)

Cartridge 7.62x39 Soviet M43 rimless

Operation Gas-operated, selective fire

Length 874mm

Weight 3.8kg, with empty magazine

Barrel 415mm, four grooves, right-hand twist

Magazine Detachable thirty-round box

Rate of fire 650±30rds/min

Selector markings In Chinese (military)

Muzzle velocity 710m/sec

Bayonet Integral triangular-blade folding pattern

The rear sling swivel sits on top of the butt wrist. Most of the receiver is made from stampings, but the bolt cover lacks strengthening ribs. An 800m back sight is retained. Ideographs appear with the selector lever, although export versions may be marked 'L' and 'D'. Machining marks are evident, minor parts are crude,

This gun replaced the original Type 56 (q.v.), although the designation remained unchanged. Bolt covers are usually AK-pattern with a plain surface instead of the ribbed version associated with the Soviet AKM; the folding bayonet may be absent. A typical gun has a wood butt and fore-end, and a plain wooden pistol grip.

fitting is poor, and a curious semi-matt finish is used.

The factory identifier (eg '66' in a triangle) generally appears on the left side of the receiver, ahead of the designation '56', the Chinese ideograph for 'Type' and the serial number, such as '20082767'. The last five digits of the number are repeated on the bolt cover.

Chinese Type 56 rifles (all types) may be found with a Type 67 muzzle attachment, which allows them to fire 70mm anti-tank grenades.

In the 1970s, production was switched from Factory No. 66 to a plant under the supervision of the China North Industries Corporation. These guns are often marked 'M22' and may have the prefix 'N' in their serial numbers. The quality of Norinco Kalashnikovs is noticeably better than Factory No. 66's military output.

Similar guns

Type 56-1 (AKM type) Essentially, this is similar to the Type 56, except for its overall length (700mm, butt folded). The struts of the butt have two indistinct spot welds on each side, behind the pivot, and the swivel may be on the top edge of the butt instead of the left side. Many of these guns have been made without bayonets, which are of little use to troops for whom small size and weight are important considerations.

Markings on a typical rifle include

ideographs accompanying the selector. The factory mark (eg '66' in a triangle) is stamped ahead of the designation '56-1 Type' and the serial number; the last five digits of the number are repeated across the back of the receiver cover.

7.62mm Type 56-2 assault rifle

Made by State Factory No. 66 and an unidentified Norinco plant.

Specification

Data from manufacturer's figures

Calibre 7.62mm (.30)

Cartridge 7.62x39 Soviet M43 rimless

Operation Gas-operated, selective fire

Length 874mm (butt extended), 654mm (butt folded)

Weight 3.9kg, with loaded magazine

Barrel 415mm, four grooves, right-hand twist

Magazine Detachable thirty-round box

Rate of fire 650±30rds/min

Selector markings In Chinese (military)

Muzzle velocity 710m/sec

Bayonet Detachable folding pattern (optional)

Introduced in the mid-1970s, the Type 56-2 rifle has a metal skeleton stock that folds to the right against the side of the receiver. A red-brown plastic cheek piece provides an important distinguishing feature. Some guns of this type have been seen with tubular butts inspired by the Belgian FNC.

Presumably, these date from the 1980s, although information is lacking.

Similar guns

Type 56-S (often mistakenly identified as Type 56-5) Advertised in the mid-1980s, this seems to be a version of the Norinco Type 56-2 that is restricted to semi-automatic fire and was originally intended for the US market.

7.62mm Type 56-C assault rifle

Made by an unidentified Norinco plant.

Specification

Data from manufacturer's figures (1996)

Calibre 7.62mm (.30)

Cartridge 7.62x39 Soviet M43 rimless

Operation Gas-operated, selective fire

Length 765mm (butt extended), 563mm (butt folded)

Weight 3.5kg, with empty magazine

Barrel 345mm, four grooves, right-hand twist

Magazine Detachable thirty-round box

Rate of fire 700±50rds/min

Selector markings In Chinese (military)

Muzzle velocity 700m/sec

Bayonet Optional

This is a compact version of the standard Type 56, with plastic furniture and a side-folding butt; a cleaning kit is carried inside the hollow pistol grip. Although the muzzle-brake and sights vary slightly in detail, internally the operating mechanism remains largely unchanged.

Egypt/United Arab Republic

7.62mm Misr assault rifle
Made by Factory No. 54 of Maadi Military & Civil Industries Company.

Specification
Generally as Soviet AKM (q.v.), except selector and sight markings in Arabic

The Egyptian armed forces have used Soviet AK and AKM rifles in addition to several million AKM-type Misr rifles. The last were made in a factory equipped with Soviet assistance in the late 1960s. The Misr has a laminated wood butt and fore-end, and a chequered plastic pistol grip. In the 1970s, guns of this type were sold in the USA by Steyr-Daimler-Puch of America Inc, under a variety of names and designations, including 'ARM'. When intended for sporting or security use, the North American versions are usually restricted to semi-automatic fire and have ten-round magazines.

Finland

The first prototypes of the improved Kalashnikov were made in the mid-

Egyptian soldiers, pictured during the Yom Kippur War (1973), with Kalashnikov rifles. A 9mm Port Said submachine-gun is slung over the shoulder of the man on the right.

1950s, and the m/58 Valmet prototype provided the basis for the perfected designs. In addition to Sako- and Valmet-made guns, however, Finland has employed the standard Soviet AK as the *Rynnäkkokiväär m/54*. The folding-stock version was used by paratroops and police, but all Soviet guns were withdrawn in the mid-1980s.

In 1960, field trials were undertaken with 200 Type A rifles provided by Sako and 200 Valmet-made Type B versions. The Sako pattern lacked a conventional trigger guard, relying on a vertical post between the trigger lever and the magazine. It had a fluted fore-end pierced with ventilation holes. The receiver was a machined forging, selec-

tors being marked 'S' and 'Y', and a bar in the butt provided a sling anchor point. The trigger guard of the Valmet gun was released by a spring-catch on the back of the pistol grip, and the butt tubes were usually covered with rubberised insulation. The Valmet factory mark appeared on the left side of the receiver. Among the experimental bayonets were a few with Beretta-type folding blades.

7.62mm Model 60 assault rifle

Rynnäkkokiväär fm/60 (RK fm/60)
Made by Valmet Oy, Jyväskylä.

Specification
Calibre 7.62mm (.30)
Cartridge 7.62x39 Soviet M43 rimless

Operation Gas-operated, selective fire
Length 915mm
Weight 4.1kg, with empty magazine
Barrel 420mm, four grooves, right-hand twist
Magazine Detachable thirty-round box
Rate of fire 750±25rds/min
Selector markings '•••' above '•'
Muzzle velocity 710m/sec
Bayonet Detachable knife pattern

The experimental Valmet Type B (or fm/58) guns were preferred to their Sako equivalents, and 200 fm/60 versions were delivered for field trials in the winter of 1960. Each had a plastic fore-end, a tubular steel butt and a back sight on the receiver cover.

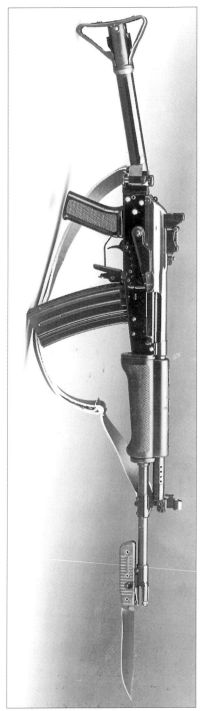

The Finnish 7.62mm M62/76 TP assault rifle with its distinctive bayonet.

A Finnish soldier on exercise fires his M62 assault rifle.

7.62mm Model 62 assault rifle

Rynnakkokivääri m/62 (RK m/62)

Made by Valmet Oy, Jyväskylä, and Oy Sako Ab, Riihimäki.

Specification

Calibre 7.62mm (.30)

Cartridge 7.62x39 Soviet M43 rimless

Operation Gas-operated, selective fire

Length 915mm

Weight 4.0kg, without magazine

Barrel 420mm, four grooves, right-hand twist

Magazine Detachable thirty-round box

Rate of fire 750±25rds/min

Selector markings '•••' above '•'

Muzzle velocity 710m/sec

Bayonet Detachable knife pattern

This rifle was derived from the semi-experimental fm/60. It has a simplified receiver and modified ribbed plastic hand guard and pistol grip. The furniture has a dark greenish hue on the original Sako-made examples, but is black on the later Valmet guns. The gas tube of the original gun lies in a stamped liner, with the top

The M71 Valmet Kalashnikov differs from the standard Finnish service issue in the design of its sights. These are mounted at the muzzle and over the chamber, in similar fashion to Soviet rifles.

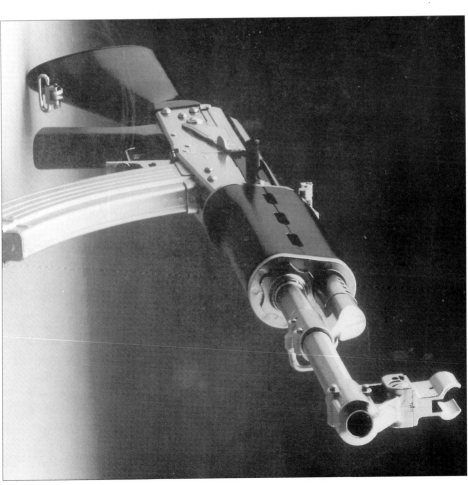

exposed, but is enclosed on later versions. The back sight attachment was improved in the late 1960s, when a solid-top front sight hood replaced the previous open pattern.

Selector dots are impressed on early guns, but raised on later examples. The serial numbers of Valmet-made guns begin at 100001, whereas Sako products commence at 200001.

From 1972, Tritium night sights were fitted to new guns, and the rounded back sight protectors were replaced by taller square versions. Rifles with the original sights were reclassified as m/62 PT.

After purchasing small numbers of

M62/76 guns, the Finns returned to the folding-butt M62 in 1986, as the stamped-steel receiver was less durable than the original machined forging. Orders for about 40,000 new guns were placed with Sako-Valmet, but apart from dated markings and minor improvements in the action and sights, these are identical to the original M62.

Similar guns

M62/S A few semi-automatic rifles were sold commercially. Usually, they have fixed or folding tubular butts, although guns touted in the USA often have conventional fixed wooden butts and wooden pistol grips.

7.62mm Model 71 assault rifle
Rynnakkokiväri m/71 (RK m/71)
Made by Valmet Oy, Jyväskylä.

Specification
Generally as m/62

This is a short-lived variant of the M62. It has a stamped receiver and a front sight assembly that is much more like the muzzle-mounted Soviet AKM pattern than the gas-port type of the Finnish service rifle. The synthetic butt, similar to the Fabrique Nationale CAL pattern, is accompanied by a parallel-sided fore-end that envelopes the gas cylinder. Most

A fixed-butt example of the Valmet M76 Kalashnikov rifle.

The 7.62mm Valmet M76 assault rifle on army exercise.

guns have a mechanical hold-open, while the back sight is positioned ahead of the ejection port instead of on the bolt cover.

Similar guns

M71S This is a semi-automatic version, offered in 5.56x45 and 7.62x39.

7.62mm Model 76 assault rifle
Rynnakkokivääri m/62/76 (RK m/62/76)
Made by Valmet Oy and
Sako-Valmet, Jyväskylä.

Specification
Data for m/62/76 TP
Calibre 7.62mm (.30)
Cartridge 7.62x39 Soviet M43 rimless
Operation Gas-operated, selective fire
Length 950mm (butt extended),
710mm (butt folded)
Weight 4.04kg, with empty magazine
Barrel 420mm, four grooves,
right-hand twist
Magazine Detachable thirty-round box
Rate of fire 750±25rds/min
Selector markings '•••' above '•'

Muzzle velocity 710m/sec
Bayonet Detachable knife pattern

Introduced in 1977, the M76 rifle may be identified by its sheet-steel receiver. Guns were purchased by the Finnish armed forces in small numbers, but it has not proved durable enough to displace the M62. Some have been sold to Qatar and others to Indonesia. The Finnish m/62/76 fixed-butt rifle has a conventional three-position selector lever on the right side of the receiver;

The Valmet M78 light machine-gun — derived from the standard M76 assault rifle — has a heavier barrel, a bipod, and Soviet-influenced sights.

m/62/76 TP signifies a folding butt model, or *Taittoperä*.

A three-piece cleaning rod, cleaning kit, screwdriver, oil bottle and brush can be carried in the butt-tube, being inserted through a pivoting cover.

On 1 January 1987, Sako and Valmet amalgamated as Sako-Valmet Oy. After several changes in the controlling interests, the Sako name was bestowed on a privately capitalised company that continued to manufacture rifles and ammunition. This led to the development of the so-called m/90 and m/92 rifles, and then to the M95 as described on page 82.

Similar guns

M76F This designation applies to the standard commercial version of the rifle with selective-fire capability and a tubular butt that folds to the right.

M76P The standard selective-fire version, this rifle has a fixed plastic butt.

M76T Distinguished by a fixed tubular butt, this version of the selective-fire Finnish Kalashnikov is mechanically similar to the other guns in the series.

M76W The suffix indicates that this gun has a fixed wooden butt.

Law Enforcement Series These are minor variants of the standard commercial weapons — F, P, T and W

versions — which have their trigger mechanisms adapted to prevent fully-automatic operation.

7.62mm Model 76B rifle

Also known as Model 82

Made by Valmet Oy, Jyväskylä.

Specification
Calibre 7.62mm (.30)
Cartridge 7.62x39 Soviet M43 rimless
Operation Gas-operated, selective fire
Length 710mm
Weight 3.315kg, without magazine
Barrel 420mm, four grooves, right-hand twist
Magazine Detachable thirty-round box

The experimental Sako M90 Kalashnikov-type rifle. Note the design of the folding butt, the back sight and the transparent magazine.

7.62mm Model 78 light machine-gun
Made by Valmet Oy, Jyväskylä.

Specification
Calibre 7.62mm (.30)
Cartridge 7.62x39 Soviet M43 rimless
Operation Gas-operated, selective fire
Length 1060mm
Weight 6.01kg, with loaded thirty-round magazine
Barrel 570mm, four grooves, right-hand twist
Magazine Detachable box: fifteen- or thirty-round box,

or 75-round drum
Rate of fire 650±50rds/min
Selector markings '●●●' above '●'
Muzzle velocity 745m/sec
Bayonet None

This is a straightforward derivative of the M76 assault rifle, having a heavy barrel and a bipod. The sights are placed in the Soviet fashion — on the receiver above the chamber and at the muzzle — and two ventilation slots are cut through the short wooden hand guard. A folding carrying handle is

Rate of fire 775±50rds/min
Selector markings '●●●' above '●'
Muzzle velocity 710m/sec
Bayonet Detachable knife pattern

This 'bullpup' version of the M76 (touted in 1981-2) comprises a standard action inserted into a one-piece synthetic stock, which reduces overall length considerably. The trigger and pistol grip are in front of the magazine, and a special raised back sight is fitted ahead of the ejection port. However, the M76B handled oddly and was never made in quantity.

fitted, and the wooden butt has a deep-belly pattern inspired by the Soviet RPD light machine-gun.

Among the optional extras are a bayonet, a blank-firing attachment and a bolt hold-open. Some 7.62x39 guns also have been fitted with the recoil buffer associated with the rarer 7.62x51 type.

Similar guns

M78S This is a heavy-barrel 'Law Enforcement' rifle, restricted to semi-automatic fire, but structurally identical to the light machine-gun.

7.62mm Model 90 assault rifle

Rynnakkokivääri m/90 PT (RK m/90 PT)
Made by Sako Oy, Jyväskylä.

Specification
Data from manufacturer's literature
Calibre 7.62mm (.30)
Cartridge 7.62x39 Soviet M43 rimless
Operation Gas-operated, selective fire
Length 930mm (butt extended), 675mm (butt folded)
Weight 3.85kg, with empty magazine, but minus accessories
Barrel 416mm, four grooves, right-hand twist
Magazine Detachable thirty-round box
Rate of fire 700±50rds/min
Selector markings '•••' above '•'
Muzzle velocity 710m/sec
Bayonet Optional

Development of this rifle began in 1986, although the first prototypes dated from 1989-90. Considerable changes were made to the selector, which was moved to the left side of the receiver where, ironically, it had been placed on the Kalashnikov prototypes of 1946. A spring-loaded cover plate was added to prevent snow and dust from entering the charging-handle slot, and many other improvements were made internally.

A transparent magazine allowed the firer to check the state of loading at a glance, changes were made to the back sight and its eared protectors, and a sling loop was added beneath the gas-port/front sight assembly. The butt was made from two sturdy steel tubes, and an improved muzzle-brake/flash hider was developed. Among the optional extras available were a bipod, optical and electro-optical sights, a grenade launcher, a blank-firing attachment and a knife bayonet.

Unfortunately, the alterations were too radical, and the Finnish Army accepted a simpler m/92/62 pattern. This became the m/95 TP described below.

7.62mm Model 95 assault rifle

Rynnakkokivääri m/95 TP (RK m/95 TP)
Made by Sako Oy, Jyväskylä.

Specification
Calibre 7.62mm (.30)
Cartridge 7.62x39 Soviet M43 rimless
Operation Gas-operated, selective fire
Length 935mm
Weight 4.5kg, with loaded magazine
Barrel 420mm, four grooves, right-hand twist
Magazine Detachable thirty-round box
Rate of fire 660±25rds/min
Selector markings '•••' and '•'
Muzzle velocity 710m/sec
Bayonet Detachable knife pattern

Adopted instead of Sako's m/90 (q.v.), the m/95 TP (*Taittoperä* — folding stock) has a tubular butt that can be swung forward to lie alongside the receiver. A modified magazine housing provides extra support for the magazine, and a rail for optical and electro-optical sights is fitted to the left side of the receiver. The charging handle is angled upward so that, if preferred, it can be retracted with the left hand. The gun has additional open Tritium dusk sights, while an improved brake/compensator is attached to the muzzle.

In 1998, assault rifle production came to an end in Finland, ensuring that the m/95 derivative of the Kalashnikov will never be made in quantity.

Similar guns

M95-S This is a semi-automatic sporting gun with a fixed wooden butt instead of the folding metal pattern.

German Democratic Republic

7.62mm MPi-K assault rifle

Maschinenpistole Kalashnikow

Made by VEB Fahrzeug- und Waffenfabrik 'Ernst Thälmann', Suhl.

Specification

Generally as Soviet AK, except selector markings ('D' above 'E')

Made in the former Sauer factory in Suhl, these guns have wooden butts, fore-ends

and hand guards, but lack the cleaning rod and butt-trap of their Soviet prototypes. A cleaning rod is carried beneath the barrel, and an 800m back sight is used.

German-made Kalashnikovs can be identified by marks such as '64 S 6511' on the left-hand side of the receiver. The first two digits represent the date of manufacture (*eg* '64' for 1964). The number 'S 6511' is also found across the back of the bolt cover. An inspector's mark, consisting of a letter inside a circle of dots, is also found.

Most German guns are made without bayonet lugs.

Similar guns

MPi-KS (*Maschinenpistole Kalashnikow mit geklappbarem Schulterstütze*) This has wooden furniture and a plain-sided folding butt, but otherwise is similar to the basic MPi-K. A typical 1961-vintage gun will bear the date and serial number (*eg* '61 Z 2691') on the left side of the receiver, the prefixed number being repeated on the back of the bolt cover, beneath the release latch. The gun will

This typical MPi-KM displays the distinctive stippled synthetic furniture associated with guns made in the German Democratic Republic.

also carry inspectors' marks in the form of small letters and encircled numbers, and will have a large proof (or perhaps quality control) mark in the form of the letter 'U' surrounded by eight dots.

7.62mm MPi-KM assault rifle

Made by VEB Fahrzeug- und Waffenfabrik 'Ernst Thälmann', Suhl.

Specification

Generally as Soviet AKM, except selector markings ('D' above 'E').

The earliest rifles of this type have wooden furniture, but later examples are fitted with plastic hand guards, wooden fore-ends and plastic butts with a distinctive stippled finish. Most guns have a cleaning rod and an 800m back sight.

Although blue-grey plastic fittings are common, some guns may have chocolate-brown versions. A typical gun will display the dated serial number (eg '67 E 2766) on the left side of the receiver, the letter suffix being repeated above the number on the back of the bolt cover, above the release catch. Inspectors' marks take the form of letter/numeral combinations (eg 'K 3') within small ovals.

Two views of a typical Hungarian AK, which has a ribbed bolt cover. Note the position of the sling swivel on the receiver side and the absence of a bayonet lug.

Similar guns

MPi-KMS This is a minor variant of the MPi-KM with a rod-type butt, which can be swung to the right once the thumb-latch protruding from the receiver has been pressed downward. The stamped-steel shoulder piece, which incorporates two distinctively slotted triangular webs, is retained by two rivets.

Combination date/serial marks, such as '71 J 2239', appear on the left side of the receiver, 'J' above '2239' being repeated across the back of the bolt cover. Many guns of this type were supplied to the Middle East and will be encountered with back sights graduated in Arabic.

MPi-KMS 72 This is a minor adaptation of the original MPi-KMS. Instead of the slotted pressed-steel shoulder plate of the original butt, it has a one-piece pattern made by turning a rod back on itself until the tip could be attached by means of spot welding. This arrangement provides a skeletal shoulder plate in the form of a flattened triangle.

Hungary

7.62mm AK-55 assault rifle
Automat Kalashnikov AK-55

Made by Fegyver é Gázkészülékgyár (FÉG), Budapest.

Two examples of the AMD-65, a short-barrel version of the AKM with a unique ventilated, sheet-metal fore-end and an auxiliary pistol grip. Note that one gun lacks its muzzle-brake and has a short twenty-round magazine.

Specification
Generally as Soviet AK, except selector markings ('∞' above '1')

The Hungarian-made version of the Soviet AK displays surprisingly good manufacturing quality, with hardwood furniture and chequering on the pistol grip. The guns have cleaning rods, but usually lack bayonet lugs. Occasionally, they may be found with white-metal or alloy blank-firing adaptors.

A typical gun may be distinguished by its selector markings, and by the presence of a serial number, such as 'AC 4425', in the machined depression on the left side of the receiver. Usually, this is repeated across the back of the bolt cover, beneath the release catch. Inspectors' marks in the form of small letters, often encircled, may also appear; some may even take the form of a small state emblem above a number.

7.62mm AKM-63 assault rifle
Made by Fegyver é Gázkészülékgyár (FÉG), Budapest.

Specification
Generally as Soviet AKM, except selector markings ('∞' above '1')

Introduced in 1963 and made until the late 1980s, this FÉG product, initially offered with a wooden butt, has a unique metal

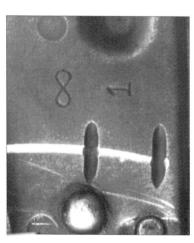

The distinctive Hungarian selector markings include '∞' for automatic fire.

fore-end formed integrally with the receiver. A supplementary pistol grip is fitted ahead of the magazine. Early examples have a pale grey-blue butt and pistol grip, but these will be either dark green or grey-black polypropylene on later guns.

7.62mm AMD-65 submachine-gun
Made by Fegyver é Gázkészülékgyár (FÉG), Budapest.

Specification
Calibre 7.62mm (.30)
Cartridge 7.62x39 Soviet M43 rimless
Operation Gas-operated, selective fire

Length 825mm (butt extended), 598mm (butt folded)
Weight 4.1kg, with empty magazine
Barrel 320mm, four grooves, right-hand twist
Magazine Detachable thirty-round box
Rate of fire 600±25rds/min
Selector markings '∞' above '1'
Muzzle velocity 690m/sec
Bayonet None

A shortened derivative of the AKM-63, the AMD has a simple tubular butt and a composite steel/rubber butt plate. The butt can be folded by pressing upward on a slotted-head catch that protrudes from beneath the receiver, behind the rear pistol grip. Both pistol grips are grey-green plastic mouldings. The short barrel is accompanied by a shortened gas-piston system, and normally a distinctive two-port muzzle-brake/compensator is fitted.

Typically, markings are confined to serial numbers (eg 'DB 4266') on the left side of the receiver and on the back of the bolt cover, beneath the release catch. Inspectors' marks usually comprise small letters or numerals within circles.

Similar guns
Some AMD rifles have been converted to fire grenades, acquiring a launcher on the muzzle and a special optical sight on a mounting plate attached to the receiver. With these, the gas valve has to be closed

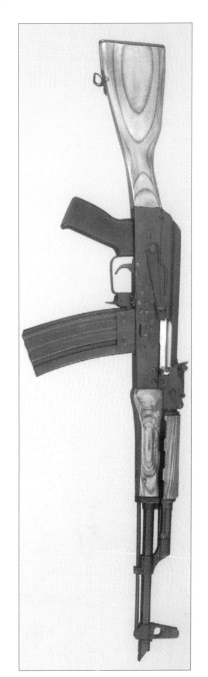

The Iraqi Tabuk assault rifle is a straightforward copy of the AKM.

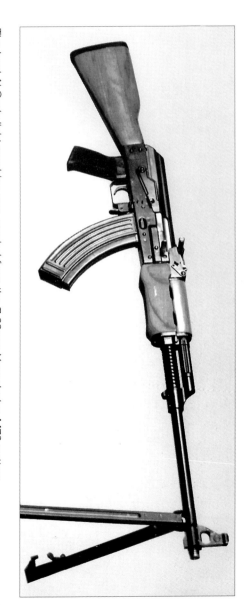

The Iraqi Al Quds light machine-gun, copied from the 7.62mm Yugoslavian M72 pattern.

before projecting a grenade, isolating the piston to allow the gun to be used virtually as a single-shot, bolt-action rifle. A shock absorber is incorporated in the tubular butt, and a new plastic fore-end — with ventilation slots in the underside — is allowed to slide backward under the control of a sturdy spring.

Iraq

7.62mm Tabuk assault rifle
Made by the State Rifle Factory.

Specification
Generally as Soviet AKM (q.v.)

The selector markings of the North Korean Type 58 assault rifle.

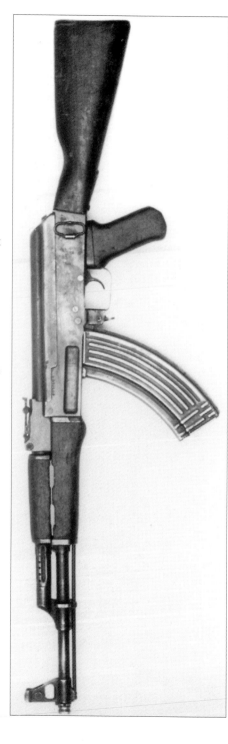

A North Korean Type 58 (AK) assault rifle. These guns are quite crudely made and can be identified by their selector markings.

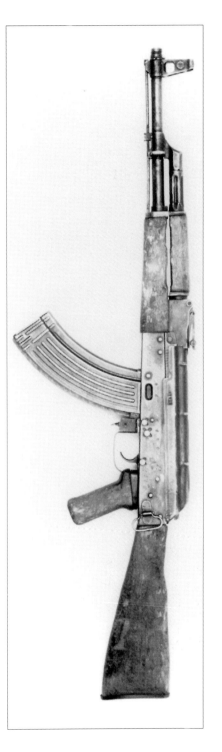

The North Korean Type 68 assault rifle was copied from the AKM.

This Kalashnikov copy is difficult to distinguish from its prototype, although the shapes of the butt and fore-end differ. Normally, the selector and sight markings are in Arabic, but Tabuk assault rifles have been made for export and marked appropriately.

Apparently, some examples of the Tabuk have also been offered with 5.56x45 chambering.

7.62mm Al Quds light machine-gun
Made by the State Rifle Factory.

Specification
Generally as Yugoslavian M72B1 (q.v.)

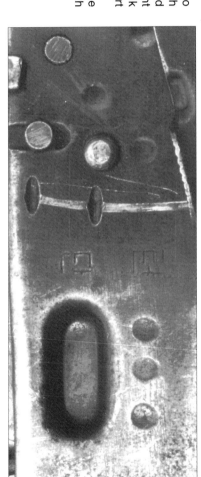

The selector markings of the North Korean Type 68 Kalashnikov copy are noticeably different from the marks of the earlier Type 58.

number is repeated across the back of the bolt cover, beneath the release catch. Tiny inspectors' marks, in the form of encircled Korean characters, will also be found on various components.

Similar guns

A folding-butt version of the Type 58 has been made in small numbers. Its Korean designation remains unknown.

7.62mm Type 68 assault rifle
Made by State Factory No. 65.

Specification
Generally as Chinese Type 56 (AKM pattern), except selector markings in Korean script

This is a modified form of the AKM, developed to suit Korean manufacturing facilities. The style of the selector markings differs noticeably from those on the earlier Type 58 rifle, and the designation is clearly marked on the front left side of the receiver.

Similar guns

The Type 68 has also been made with a distinctive folding butt. This has four slots in each side bar to conserve weight.

7.62mm TUL-1 light machine-gun
Manufacturer unknown.

National and designation markings on a North Korean Type 68 (AKM) rifle.

This is simply a licence-built Kalashnikov light machine-gun, which may be distinguished from its Yugoslav prototype largely by inferior manufacturing standards. Its sights and selectors are stamped in Arabic, although some export models have been marked differently.

People's Republic of Korea

7.62mm Type 58 assault rifle
Made by State Factories No. 61 and No. 65.

Specification
Generally similar to the Soviet AK, except

selector markings in Korean script

Apparently, the North Koreans originally used Chinese Type 56 and 56-1 (q.v.) Kalashnikov rifles, but eventually began to make guns of their own. These have wooden furniture, often with laminated pistol grips, and are characterised by a very crude standard of manufacture with prominent tool marks. They have cleaning rods, but lack bayonet lugs.

Typical markings comprise '✪' ahead of a serial number that consists of two Korean characters and a five-digit number, such as '12480'. Usually, the

Specification
Calibre 7.62mm (.30)
Cartridge 7.62x39 Soviet M43 rimless
Operation Gas-operated, selective fire
Length 1050mm (approximately)
Weight 5.5kg (approximately)
Barrel 590mm, four grooves, right-hand twist
Magazine Detachable thirty-round box
Rate of fire 650±30rds/min
Selector markings In Korean script
Muzzle velocity 735m/sec
Bayonet None

Basically, this is a modified AK-type receiver with the bipod and barrel of the RPD (Degtyarev) light machine-gun. The quality of the work suggests that the alterations were undertaken to meet local needs and were not carried out by one of the major North Korean manufacturers.

Poland

7.62mm PMK assault rifle
Pistolet Maszynowy Kalasznikow (PMK)
Made by Zaklady Metalowe Lucznik, Radom.

Specification
Generally as Soviet AK, except selector markings ('C' above 'P') and lack of bayonet lug

The Polish version of the AKMS, shown here on exercise in 1981.

Production of this gun, also known as the *Karabinek Automatyczny Kalasznikow (Kbk-AK)*, began in the early 1960s. Essentially, the PMK is similar to the original Soviet gun, with wooden furniture, but has a chequered pistol grip of a different form. It has a cleaning rod and normally carries an 11-in-oval factory mark, the date and a prefixed serial number (eg '1964' and 'MJ12891'). Usually, the five digits of the serial number are repeated on the back of the bolt cover, beneath the release catch. Inspectors' marks may also be found, in the form of letters or numbers within tiny diamonds. Manufacturing quality is surprisingly good.

Similar guns

PMK-DGN Also known as the Kbk-G wz/60, this variant was adapted to fire F1/N60 anti-personnel and PGN-60 anti-tank grenades from the LON-1 launcher. It has a special leaf sight, a ten-round magazine and a gas-cylinder cut-off valve marked 'O' (open) and 'Z' (closed). Trapezoidal metal plates containing small recesses, on each side of the butt, retain the straps of a boot-type recoil pad.

A typical gun will have the factory identifier ('11' in an oval) above the year of production and the serial number (eg 'KZ05197'). The prefixed number is repeated in full on the fore-end, ahead of the receiver, and minus the prefix on the back of the bolt cover, beneath

the release catch; the gas-valve lever will also be marked with the number. Inspectors' marks take the form of diamonds containing letter or number/letter combinations.

PMK-S Essentially similar to the standard PMK, this rifle has a C-section folding butt with a U-type shoulder piece. The pistol grip and the plain fore-end are wooden. A typical gun will bear a factory mark of '11' in an oval on the right side of the receiver, above the year and the serial number (eg 'MK 12313'). The digits of the serial number will be repeated across the back of the bolt cover, beneath the release catch.

The current version of the Polish AKMS.

7.62mm PMK-M assault rifle

Made by Zaklady Metalowe
Lucznik, Radom.

Specification

Generally as Soviet AKM (q.v.), except
selector markings ('C' above 'P')

Production of an AKM-type rifle began in
the mid-1960s in Factory No. 11. It has a
laminated fore-end, a laminated hand
guard and a chequered plastic pistol grip.
Manufacturing quality is quite good: the
stamped receiver and bolt cover, for

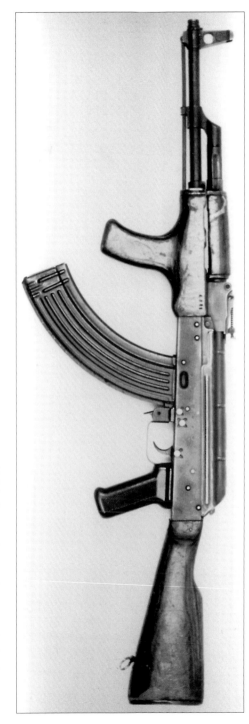

example, together with most of the minor
external parts, are usually well blacked. A
cleaning rod is carried beneath the
muzzle, and the lower part of the short
muzzle attachment is extended to serve
as a rudimentary compensator.

Similar guns

PMK-MS This is identical to the PMK-M,
except for a folding stock with three rivets,
three spot welds and two short flutes on
each side of the pressed-steel strut. The
fore-end and hand guard are usually of
wood laminate, but the pistol grip is

chequered plastic.

7.62mm Radom Hunter rifle

Made by Zaklady Metalowe
Lucznik, Radom.

Specification

Mechanically similar to Polish AKM copy.

Intended for commercial sale, this rifle is
simply a 7.62x39 AKM action (restricted
to semi-automatic fire) in a thumbhole-
style wooden stock. It does not seem to
have been made in large numbers, but

*A Romanian AIM, based on the AKM. This gun is easily distinguished by its fore-end, which is formed into a pistol grip. It also has
distinctive selector markings; in the case of this particular gun, '1', '2' and '3'.*

surprisingly has been widely distributed in Germany.

Romania

7.62mm AI assault rifle

Made in the state arms factory, Cugir.

Specification
Generally as Soviet AK (q.v.), except selector markings ('S' above 'FA' above 'FF').
This rifle is little more than a copy of the Soviet AK, sharing similar construction. Unlike the later Romanian AIM (q.v.), the AI is rarely found with the distinctive auxiliary pistol grip.

Similar guns

A folding-butt version of the AI was introduced in the 1970s, followed by a light machine-gun in about 1983. The latter is essentially similar to the RPK, but features a different method of attaching the bipod.

7.62mm AIM assault rifle

Made in the state arms factory, Cugir.

Specification
Calibre 7.62mm (.30)
Cartridge 7.62x39 Soviet M43 rimless
Operation Gas-operated, selective fire
Length 880mm
Weight 4.0kg, without magazine
Barrel 415mm, four grooves, right-hand twist
Magazine Detachable thirty-round box
Rate of fire 650±30rds/min
Selector markings 'S' above 'FA' above 'FF'
Muzzle velocity 710m/sec
Bayonet Detachable knife pattern

Derived from the Soviet AKM, the improved AIM rifle has a wooden butt and wood-laminate hand guard and fore-end, which has an integral pistol grip. The rear pistol grip is injection-moulded plastic with chequered panels. The bolt cover is a ribbed pressing, there is a 1000m back sight, and a cleaning rod is carried beneath the barrel.

The bayonet is attached by the customary lug. Normally, the rear sling swivel is fitted to the under-edge of the butt, although many guns have been found with the swivel moved to the left side of the butt-wrist, probably as a result of field modifications by insurgents who have carried guns around their necks for long distances.

Generally, Romanian Kalashnikovs bear an arrowhead-in-triangle mark on the left side of the receiver, ahead of a date/serial number group, such as '1967 A10184'. The prefixed number is duplicated across the back of the bolt cover. Inspectors' marks usually consist of small encircled letters, or a small triangle containing an arrowhead above a one- or two-digit number. The selectors of guns intended for export may be marked '1', '2' and '3'.

7.62mm AIR assault rifle

Made in the state arms factory, Cugir.

Specification
Calibre 7.62mm (.30)
Cartridge 7.62x39 Soviet M43 rimless
Operation Gas-operated, selective fire
Length 750mm (butt extended), 550mm (butt folded)
Weight 3.1kg, without magazine
Barrel 305mm, four grooves, right-hand twist
Magazine Detachable box: twenty or thirty rounds
Rate of fire 600±30rds/min
Selector markings 'S' above 'FA' above 'FF'
Muzzle velocity 670m/sec
Bayonet None

This is a compact version of the AIM, with a short barrel and a single-strut butt that folds sideways to reduce overall length. Most guns lack the pistol grip fore-end that is characteristic of Romanian Kalashnikovs, but apparently this is available as an optional extra.

The AK-103 is currently being made by Izhmash.

Russia

7.62mm AK-103 assault rifle

Avtomat Kalashnikova obr. 103 (AK-103)
Made by Izhmash AO, Izhevsk,
Udmurt Republic.

Specification
Calibre 7.62mm (.30)
Cartridge 7.62x39 Soviet M43 rimless

Operation Gas-operated, selective fire
Length 943mm (butt extended), 700mm
(butt folded)
Weight 3.4kg, without magazine
Barrel 415mm, four grooves,
right-hand twist
Magazine Detachable thirty-round box
Rate of fire 775±25rds/min
Selector markings 'AB' over 'ОД'
Muzzle velocity 715m/sec
Bayonet Detachable knife pattern

A minor derivative of the AK-74, the AK-103 has plastic furniture and a butt that can be folded forward along the left side of the receiver. The butt is grooved on the right side and has a swivel on the wrist. The plastic pistol grip has a chequered panel, while the fore-end is extensively ribbed and has a large rounded ledge to improve hand grip. The back sight is a 1000m type, and a rail for optical or electro-optical sights is fitted to the left

95

The AK-104, currently in production, is little more than a short-barrelled version of the AK-103.

side of the receiver.

Changes have been made to the shape of the gas-port assembly, which is noticeably squarer than the original AK/AKM type, and the long cylindrical muzzle-brake/compensator has also been improved. Two lugs allow the optional tool-bayonet to be attached, even if the compensator has been removed. Standard accessories include a sling, an oiler and a set of cleaning tools in a small cylindrical container.

A pouch holding three spare magazines is an optional extra.

7.62mm AK-104 submachine-gun

Avtomat Kalashnikova obr. 104 (AK-104)
Made by Izhmash AO, Izhevsk, Udmurt Republic.

Specification
Calibre 7.62mm (.30)
Cartridge 7.62x39 Soviet M43 rimless

Operation Gas-operated, selective fire
Length 824mm (butt extended), 586mm (butt folded)
Weight 2.9kg, without magazine
Barrel 295mm, four grooves, right-hand twist
Magazine Detachable thirty-round box
Rate of fire 775±50rds/min
Selector markings 'AB' over 'ОД'
Muzzle velocity 670m/sec
Bayonet None

The RPK, a light-support derivative of the AKM. This gun has a forty-round box magazine, which illustrates the difficulty of maintaining a low silhouette if cover is minimal. Paint has been applied to the receiver and butt to obscure evidence of previous ownership.

Basically, this is a short-barrelled AK-103, sharing the same plastic furniture and folding butt. The muzzle-brake, which abuts the front sight/gas-port assembly, consists of a two-diameter cylinder with a short conical flash hider. A cleaning rod is carried beneath the barrel, but there is no bayonet lug. A 500m back sight and a receiver-mounted optical/electro-optical sight rail are standard fittings. The AK-104 is usually accompanied by a sling, an oiler and a small plastic container for the cleaning tools.

USSR

7.62mm RPK light machine-gun

Ruchnoi Pulemet Kalashnikova (RPK)
Made in the arms factories in Tula and Izhevsk

Specification
Calibre 7.62mm (.30)
Cartridge 7.62x39 M43 Soviet rimless
Operation Gas-operated, selective fire
Length 1040mm
Weight 5.0kg, with bipod
Barrel 590mm, four grooves, right-hand twist
Magazine Thirty- or forty-round box; seventy-five-round drum
Rate of fire 600±25rds/min
Selector markings 'AB' over 'ОД'
Muzzle velocity 735m/sec
Bayonet None

The success of the Kalashnikov assault rifle inspired the development of a comparable light support weapon, which has also been made in Hungary, Romania and the German Democratic Republic (the last as the LMG-K). The barrel of the RPK is longer and heavier than that of the AKM, and increased magazine capacity allows fire to be sustained for short periods. The muzzle velocity is greater by about 30m/sec.

A bipod is attached to the barrel, behind the front sight, and the butt duplicates the deep-bellied design of the belt-fed RPD to provide a better grip for the left hand in the prone position. The 1000m back sight can be adjusted laterally, unlike the standard rifle pattern.

Similar guns

RPKS (*Ruchnoi Pulemet Kalashnikova skladvayushimsya prikladom — РПКС* in Cyrillic) Basically, this is an RPK with a butt that can be folded to the left to reduce overall length, facilitating transport. It was developed for commando, airborne and armoured troops.

7.62mm AKMS-U submachine-gun

Avtomat Kalashnikova Modificatsioniya skladvayushimsya prikladom, ustankova obraztsa, (AKMS-U)
Made in the Tula ordnance factory.

Specification
Data supplied by Ian Hogg

Calibre 7.62mm (.30)
Cartridge 7.62x39 Soviet M43 rimless
Operation Gas-operated, selective fire
Length 722mm (butt extended)
Weight 3.35kg, with empty magazine
Barrel 225mm, four grooves, right-hand twist
Magazine Detachable thirty-round box
Rate of fire 800±30rds/min
Selector markings 'AB' above 'ОД'
Muzzle velocity 645m/sec
Bayonet None

The AKMS-U was the successful entrant in a competition to find a port-firing weapon for use in armoured personnel carriers (*samokhodnaya ustankova*); it has also been called AKR and Krinkov, apparently after the leader of the design team. Its principal competitor seems to have been a Simonov design, which may have been made in small numbers for field trials.

A compact derivative of the AKM, the AKMS-U lacks the standard fire-rate reducer system. It also has a folding U-section butt with three rivets and a single long flute on each side. Omitting the cleaning rod allows the short, finned expansion chamber on the ultra-short barrel to be fitted with a conical flash hider. The front-sight block has been moved back to abut the laminated wooden barrel guard, and a chequered plastic pistol grip is used in conjunction with a short wooden thumbhole fore-

The parts of an AK-type 7.62mm Yugoslav M70 assault rifle, from the military magazine *Front*. 1, barrel. 2, front, grenade-launching and back sights; 2a, luminescent night sights. 3, gas-port block; 3a, regulator. 4, gas-piston tube. 5, return spring. 6, piston-rod. 7, breechblock; 7a, striker. 8, top cover. 9, pistol-grip and trigger guard assembly. 10, trigger mechanism; 10a, trigger lever; 10b, disconnector; 10c, hammer; 10d, full-automatic sear; 10e, trigger tail and selector blade. 11, magazine. 12, butt. 13, cleaning rod. 14, grenade launcher.

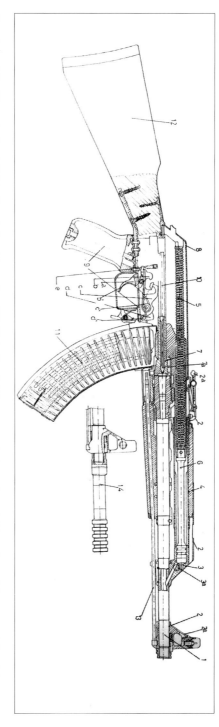

Drawing of the Yugoslavian M64A rifle, derived from the AK. Note the folding grenade-launcher sight on top of the gas-piston tube.

end. The sling attaches by means of a D-ring to a bracket on the left side of the front-sight block.

The base of the rocking-'L' back sight is combined with the receiver-cover pivot, the sight being regulated for 100m (marked 'Π') and 500m (marked '–5'). Adjustments to the point of impact can be made by rotating the front sight.

7.62mm AK-74UB submachine-gun

Manufacturer unknown.

Specification

Details unknown

This gun is said to have been converted, or derived, from the AK-74 in the late 1970s to fire a special noiseless 7.62mm SP-3 'piston seal' cartridge. It shares the general dimensions of the 5.45mm gun, but presumably is restricted to semi-automatic fire.

Yugoslavia/Serbia

7.62mm Model 64 assault rifle

Automatska puška vz. 64

Made by Zavodi Crvena Zastava, Kragujevač.

Specification

Calibre 7.62mm (.30)

Cartridge 7.62x39 Soviet M43 rimless

Operation Gas-operated, selective fire

Length 1040mm

Weight 3.9kg, without magazine

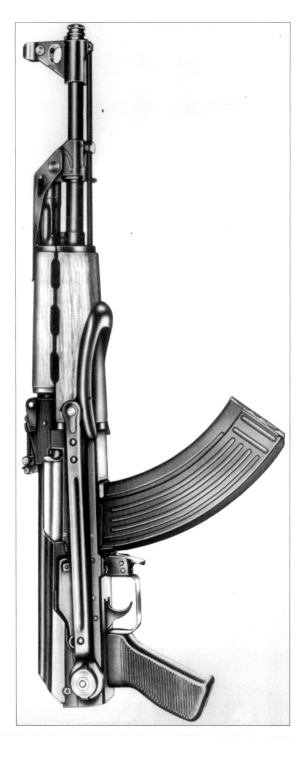

A Yugoslavian M70A Kalashnikov with a folding butt, a grenade-launcher sight and a ribbed plastic pistol grip.

Barrel 500mm, four grooves, right-hand twist
Magazine Detachable twenty-round box
Rate of fire 775±50rds/min
Selector markings 'U' above 'R' above 'J'
Muzzle velocity 730m/sec
Bayonet None

Otherwise essentially similar to the AK (q.v.), this gun introduced a long barrel, a mechanical hold-open and a special magazine (incorporating a hold-open notch), which cannot be exchanged with the Soviet thirty-round type. The Yugoslavian gun has a detachable grenade launcher on the muzzle (it can be replaced with a muzzle-brake/compensator unit) and a folding sight on top of the gas tube; the sight arm automatically cuts the gas supply to the piston as it is raised. The wooden pistol grip has prominent finger grooves, and a 1000m back sight is used.

Similar guns

M64A Derived from the original M64, this has a 415mm barrel, a thirty-round magazine and an overall length of 958mm without the grenade launcher tube. The rearward-folding grenade-launching sight is retained on top of the gas-port housing. The butt, fore-end and pistol grip of the earliest guns are all wood, but a ribbed, black plastic pistol grip will be found on later versions. Butt plates are often

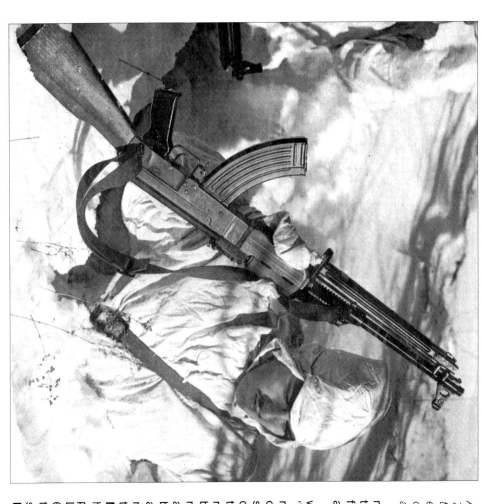

A Yugoslavian soldier, armed with a 7.62mm M65A light machine-gun, wades through waist-deep snow during an army exercise in 1977. Note the milled pattern on the receiver side, above the trigger and pistol grip.

rubber. Subsequently redesignated M70, the M64A is sturdily made, although the finish may be crude; 1000m back sights are customary.

A typical gun will be marked with 'ZASTAVA—KRAGUYEVAC' over 'YUGOSLAVIA' on the left side of the receiver, behind the distinctive concentric-circle ZCZ trademark. A serial number, such as 'C-83080' will appear in the milled depression on the front right-hand side of the 'receiver, with '83080' on the lower rear right-hand side of the receiver, behind the selector pivot. The same number will be etched lightly on the bolt, and be stamped into the left side of the butt. A mould number, (eg '3106') will appear on the top of the pistol grip, on the right side, while inspectors' marks take the form of numbers within small squares.

M64B Otherwise similar to the M64A, this has a folding butt with three rivets and a prominent longitudinal flute on each side. Length with the butt folded is a mere 690mm. Generally, the markings mirror those of the M64A. Guns of this type were subsequently redesignated M70A.

M70 This is a post-1971 designation for **101**

The Yugoslavian M70B1 assault rifle. Note the bayonet lug and the folded grenade-launcher sight on top of the gas-piston tube.

the M64A, with a few minor internal changes. It was replaced by the AKM-type Model 70B1 in about 1974.

M70A A post-1971 designation for the folding-butt M64B. This AK-type gun was soon replaced by the AKM-type Model 70AB2 (q.v.).

7.62mm Model 65A light machine-gun

Puškomitraljez vz. 65
Made by Zavodi Crvena Zastava, Kragujevač.

made as a light support weapon, with a heavy barrel and a bipod at the muzzle, behind the front-sight block. The barrel of this gun has fins between the gas-port housing and the short wooden fore-end to facilitate cooling.

Similar guns

M65B Essentially similar to the standard M65, this has a plain 'quick-change' barrel to save weight. It weighs about 5.2kg without its magazine.

7.62mm Model 70B1 assault rifle

Automatska puška vz. 70B1

The 1964-pattern assault rifle was also

Specification
Calibre 7.62mm (.30)
Cartridge 7.62x39 Soviet M43 rimless
Operation Gas-operated, selective fire
Length 1055mm
Weight 5.58kg, without magazine
Barrel 500mm, four grooves, right-hand twist
Magazine Detachable thirty-round box
Rate of fire 775±50rds/min
Selector markings 'U' above 'R' above 'J'
Muzzle velocity 735m/sec
Bayonet None.

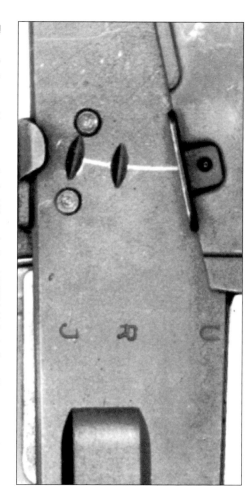

The selector markings on a typical Yugoslavian M70B1 Kalashnikov. Some guns have been seen without the 'U' marking.

Made by Zavodi Crvena
Zastava, Kragujevać.

Specification

Data from manufacturer's literature

Calibre 7.62mm (.30)

Cartridge 7.62x39 Soviet M43 rimless

Operation Gas-operated, selective fire

Length 890mm.

Weight 3.75kg, with empty magazine

Barrel 415mm, four grooves, right-hand twist

Magazine Detachable thirty-round box

Rate of fire 650±30rds/min

This gun appeared in about 1974, replacing the AK-type Model 70 (q.v.). It retains the grenade launcher and the integral sight of the latter, but the receiver is made largely from stampings and pressings. A rate reducer is built into the trigger mechanism to restrict the rate of fire.

Selector markings 'U' above 'R' above 'J'

Muzzle velocity 720m/sec

Bayonet Detachable knife pattern

Similar guns

M70B1N This is a minor adaptation of the standard gun, with a mounting rail for optical and electro-optical sights on the left side of the receiver.

M70AB2 Generally as Model 70B1, except for length — 890mm (butt extended), 640mm (butt folded) — and weight (3.7kg). A minor variant of the M70B1, replacing the original Model 70A (q.v.) this may be distinguished by its folding butt. The magazine weighs 360gm empty, and 870gm loaded.

M70AB2N This is simply a standard gun with a mounting rail for optical and electro-optical sights on the left side of the receiver.

7.62mm Model 72 light machine-gun

Puškomitraljez vz. 72

Made by Zavodi Crvena
Zastava, Kragujevać.

Specification

Calibre 7.62mm (.30)

Cartridge 7.62x39 Soviet M43 rimless

Operation Gas-operated, selective fire

Length 1025mm

Weight 5.9kg, with empty drum magazine

Barrel 540mm, four grooves, right-hand twist

Magazine Detachable thirty-round box

Rate of fire 650±30rds/min

The Yugoslavian M70AB2 has a folding butt, but otherwise is mechanically similar to the M70B1.

Selector markings 'U' above 'R' above 'J'
Muzzle velocity 748m/sec
Bayonet None

This is the first of the Yugoslavian Kalashnikov derivatives to be built on an AKM-type receiver, comprised largely of stampings and pressings. It has a reducer built into the trigger mechanism that restricts the cyclic rate, improving control when firing automatically. The fixed heavy barrel has fins between the gas-port housing and the short wooden fore-end, while the 1000m back sight has a lateral 104 adjustment on the sight-notch block. A

conventional assault-rifle butt is fitted. The M72 was short-lived, however, as it was soon replaced by the M72B1 described below.

7.62mm Model 72B1 light machine-gun
Puškomitraljez vz. 72B1
Made by Zavodi Crvena Zastava, Kragujevač.

Specification
Data from manufacturer's literature
Calibre 7.62mm (.30)
Cartridge 7.62x39 Soviet M43 rimless

Operation Gas-operated, selective fire
Length 1025mm
Weight 5.0kg, without magazine
Barrel 540mm, four grooves, right-hand twist
Magazine Detachable thirty-round box; seventy-five-round drum
Rate of fire 600±30rds/min
Selector markings 'U' above 'R' above 'J'
Muzzle velocity 745m/sec
Bayonet None

This is a heavy-barrel variant of the FAZ

The parts of a Yugoslav 7.62mm (AKM type) M72 light machine-gun, from the military magazine *Front*. 1, barrel; 1a, muzzle-nut; 2, front sight block: 2a, front sight; 2b, folding night sight; 3, back sight: 3a, lateral adjuster; 3b, sight-notch block; 4, gas-port block; 5, piston tube; 6, return spring assembly; 6a, spring; 6b guide rod; 6c, coupling; 7, bolt unit; 7a, bolt; 7b, carrier; 7c, striker; 8, top cover; 9, receiver assembly; 9a, receiver body; 9b, pistol grip; 9c, trigger guard; 9d, magazine catch; 10, trigger assembly: 10a, full-automatic sear; 10b, hammer; 10c, trigger hook; 10d, hammer-delay lever or retarder; 10e, trigger tail; 10f, selector blade; 11, magazine; 11a, follower; 11b, body; 11c spring; 11d, spring-anchor; 11e, floorplate; 12, chamber; 13, butt. 14, bipod. 15, cleaning rod.

A Yugoslavian M72AB1 light machine-gun, with its butt folded beneath the receiver. This configuration restricts the length of the magazine.

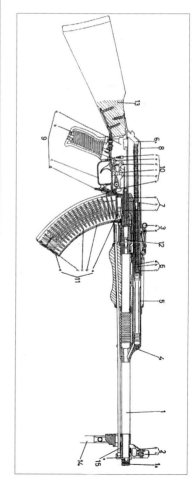

The Yugoslavian M72 light machine-gun with standard box magazine.

series intended to provide support fire, although the barrel is fixed in place, restricting its capacity to sustain fire, even though cooling fins are provided. A bipod is attached beneath the barrel, behind the front-sight block; the stock and fore-end are wooden, and a laterally-adjustable 1000m back sight is fitted.

The action is the standard Yugoslav-made AKM, with a stamped receiver. It incorporates a reducer in the trigger system to restrict the rate of fire. The drum magazine weighs 2.175kg loaded.

Similar guns

Model 72AB1 A modified version of the M72B1, this has a folding strut-type butt. The barrel remains fixed.

7.62x51

Finland

7.62mm Model 78 light machine-gun

Made by Valmet Oy, Jyväskylä.

Specification
Calibre 7.62mm (.30)
Cartridge 7.62x51 NATO rimless
Operation Gas-operated, selective fire
Length 1060mm
Weight 5.92kg, with loaded magazine
Barrel 570mm, four grooves, right-hand twist
Magazine Detachable twenty-round box
Rate of fire 650±50rds/min
Selector markings '•••' above '•'
Muzzle velocity 845m/sec
Bayonet Optional

This is a straightforward derivative of the M76 assault rifle, with a heavy barrel, a recoil buffer in the receiver, and a bipod on the barrel. The sights are on the receiver, above the chamber, and at the muzzle. Ventilation slots are cut through the short wooden hand guard; a folding carrying handle is provided immediately ahead of the back sight, and the wooden butt has a deepened belly. Optional

A 7.62x51 version of the Finnish Valmet M78 light machine-gun, which can be distinguished from the 7.62x39 version by the shape and width of its magazine.

The Finnish Petra sporting rifle is built on the standard Valmet Kalashnikov action.

An Israeli soldier, equipped with a 5.56mm Galil ARM, pauses at the entrance to a bunker in the Sinai, 1974.

extras include a bayonet, a blank-firing attachment and a bolt hold-open.

Petra sporting rifle
Made by Valmet Oy, Jyväskylä.

Specification
Calibre 7.62mm (.30)
Cartridge .308 Winchester (7.62x5l)
Operation Gas-operated, semi-automatic
Length 995mm
Weight 3.85kg, with empty magazine
Barrel 420mm, four grooves, right-hand twist
Magazine Detachable ten-round box
Rate of fire N/A
Selector markings '•' in the lower position
Muzzle velocity 800m/sec
Bayonet None

Announced commercially in 1982, and made until about 1986 in .243 and .308 chambering, the Petra is one of the earliest successful attempts at adapting the Kalashnikov for the sporting fraternity. Basically an M78 light machine-gun action, already adapted for the 7.62x51 cartridge, the Petra rifle is restricted to semi-automatic fire. The good-quality wooden butt has a straight comb, chequering on the pistol grip and a ventilated rubber recoil pad. The chequered open-top fore-end is also of wood. Swivels are fitted to the barrel and under the butt, while the receiver is equipped with an optical-sight mount.

Israel

7.62mm ARM assault rifle

Made by Israeli Military Industries,
Ramat ha-Sharon.

Specification

Data from manufacturer's literature (1984)
Calibre 7.62mm (.30)
Cartridge 7.62x51 NATO rimless

Operation Gas-operated, selective fire
Length 1050mm (butt extended),
810mm (butt folded)
Weight 4.3kg, with empty magazine,
bipod and carrying handle
Barrel 535mm, four grooves, right-hand
twist
Magazine Detachable box: twelve,
twenty-five or fifty rounds
Rate of fire 650±50rds/min

Selector markings 'S' above 'A' above 'R'
Muzzle velocity 850m/sec
Bayonet Detachable knife pattern

Introduced in 1975, this rifle is noticeably
larger and heavier than the 5.56mm gun
(q.v.), having a bulkier magazine with
virtually no curve. Some guns dating from
the 1970s have their selector-repeater
markings on the left side of the receiver,

An Israeli 7.62x51 Galil ARM fitted with the optional bipod.

Folding Stock

Receiver

Pistol Grip

Rear Sight

Rear Night Sight (folded)

Trigger Mechanism

Return Spring

Magazine Catch

Charging Handle

Carrying Handle

Bolt

Bolt Carrier

Magazine

Gas Cylinder

Gas Piston

Handguard

Front Night Sight (folded)

Front Sight and Guard

Gas Block

Folding Bipod

Barrel

Flash Suppressor

110 *The Israeli Galil rifle.*

A standard Galil AR with a short synthetic fore-end.

above the pistol grip. These read 'R' 'A' 'S' instead of the customary 'S' 'A' 'R'. However, this was not a permanent change, and it is thought that only a few rifles were affected.

7.62mm AR assault rifle

Made by Israeli Military Industries, Ramat ha-Sharon.

Specification

Generally as ARM, except for a weight of 3.95kg

A minor version of the ARM, the AR lacked its bipod and carrying handle. Apart from dimensions and the shape of

the magazine, it is essentially similar to the 5.56mm type (q.v.).

7.62mm SAR assault rifle

Made by Israeli Military Industries, Ramat ha-Sharon.

Specification

Data from manufacturer's literature (1984)

Calibre 7.62mm (.30)
Cartridge 7.62x51 NATO rimless
Operation Gas-operated, selective fire
Length 915mm (butt extended), 675mm (butt folded)
Weight 3.75kg, with empty magazine and bipod

Barrel 400mm, four grooves, right-hand twist
Magazine Detachable box: twelve, twenty-five or fifty rounds
Rate of fire 650±50rds/min
Selector markings 'S' above 'A' above 'R'
Muzzle velocity 790m/sec
Bayonet Usually none

This is a short version of the basic design, being easily identifiable by the length of the barrel and the shortened gas-piston tube. Bayonet lugs are uncommon.

7.62mm Galil Sniper rifle

Made by Israeli Military Industries, Ramat ha-Sharon.

Specification
Calibre 7.62mm (.30)
Cartridge 7.62x51 NATO rimless
Operation Gas-operated, semi-automatic
Length 1115mm (butt extended), 840mm (butt folded)
Weight 6.4kg, with empty magazine and bipod
Barrel 500mm, four grooves, right-hand twist
Magazine Detachable twenty-round box
Rate of fire N/A
Selector markings 'S' above 'R'
Muzzle velocity 850m/sec
Bayonet None

Developed for Israeli Army snipers in the early 1980s, and introduced in 1983, this special semi-automatic 7.62mm AR-type Galil is equipped with a heavyweight barrel and a large tubular muzzle-brake/compensator. A two-stage trigger is fitted, while the bipod has been moved back so that it pivots on the receiver instead of the gas block, relieving the barrel of unnecessary stress that could reduce the weapon's accuracy.

The folding wooden butt has a cheek piece and a ventilated rubber recoil pad. A two-position aperture back sight is fitted to the receiver, but a bracket attached to the left-hand side of the receiver accepts a 6x40 Nimrod sight or any NATO-standard infra-red and image-intensifying night-vision sights.

7.62mm Hadar II rifle
Made by Israeli Military Industries, Ramat ha-Sharon.

Specification
Calibre 7.62mm (.30)
Cartridge 7.62x51 NATO rimless
Operation Gas-operated, semi-automatic
Length 980mm
Weight 4.35kg, with empty magazine
Barrel 500mm, four grooves, right-hand twist
Magazine Detachable twenty-round box
Rate of fire N/A
Selector markings 'S' above 'R'
Muzzle velocity 850m/sec
Bayonet None

Intended for police use, and also occasionally sold as a sporting rifle, the 1987-vintage Hadar II has a standard AR action set into a three-quarter-length wooden stock. This has an unmistakable thumbhole butt and a radial safety lever set into the left side of the pistol grip.

Yugoslavia/Serbia

7.62mm Model 77B1 assault rifle
Automatska puška vz. 77B1
Made by Zavodi Crvena Zastava, Kragujevač.

Specification
Calibre 7.62mm (.30)
Cartridge 7.62x51 NATO rimless
Operation Gas-operated, selective fire
Length 990mm
Weight 3.4kg, without magazine
Barrel 500mm, four grooves, right-hand twist
Magazine Detachable twenty-round box
Rate of fire 600±50rds/min
Selector markings 'U' above 'R' above 'J'
Muzzle velocity 840m/sec
Bayonet Detachable knife pattern

Based on the standard M70B1 rifle, the M77B1 is equipped with a modified receiver that will accommodate the bulkier cartridge. A slotted muzzle-brake/compensator is incorporated, while the grenade-launcher tube and auxiliary ladder-type sights were reduced to the status of optional extras.

Similar guns

M77 sniper rifle Essentially similar in design to the standard Yugoslav/Serbian M77B1 assault rifle, this specialised sniper rifle features a ten-round magazine and weighs about 5.0kg. The butt and fore-end are made of wood, while the pistol grip is of ribbed plastic. Standard fittings include receiver rails for optical and electro-optical sights.

7.62mm Model 77 light machine-gun

Puškomitraljez vz. 77
Made by Zavodi Crvena
Zastava, Kragujevac̆.

Specification

Calibre 7.62mm (.30)
Cartridge 7.62x51 NATO rimless
Operation Gas-operated, selective fire
Length 1025mm

Weight 5.075kg, without magazine
Barrel 535mm, four grooves, right-hand twist
Magazine Detachable twenty-round box
Rate of fire 625±25rds/min
Selector markings 'U' above 'R' above 'J'
Muzzle velocity 840m/sec
Bayonet None

This light machine-gun is a heavy-support version of the standard Model 77B1 assault rifle, with a heavyweight barrel and a folding bipod. The barrel is fixed, however, which restricts the weapon's capability to sustain fire, particularly when using a full-power cartridge.

A Yugoslavian M77B1 light machine-gun.

7.62x54R

Romania

7.62mm FPK sniper rifle

Made in the state firearms factory, Cugir.

Specification
Calibre 7.62mm (.30)
Cartridge 7.62x54 Soviet rimmed
Operation Gas-operated, semi-automatic
Length 1110mm (approximately)
Weight 4.5kg, without magazine and sight
Barrel 500mm, four grooves, right-hand twist
Magazine Detachable ten-round box
Rate of fire N/A
Selector markings Unknown
Muzzle velocity 815m/sec
Bayonet None

Sharing the basic action of the Romanian-manufactured AIM (AKM) assault rifle, suitably elongated to accommodate the 7.62mm rimmed cartridge, the specially developed FPK sniper rifle displays a slight external similarity to the Soviet Dragunov, owing to the extended barrel, short magazine body and cutaway butt.

This gun does not seem to have been made in large numbers and is rarely seen in Western Europe.

Russia

SVD sniper rifle

Snayperskaya Vintovka Dragunova (SVD)
Made by Izhmash AO, Izhevsk Udmurt Republic.

Apart from a few minor details and a wider choice of sights, this rifle is identical to its Soviet predecessor described opposite. The standard optical sight in current use is the 4x24 PSO-1M2, which can be clamped to a rail on the left-hand side of the receiver.

Similar guns

SVDN2 The N2 has an NSPUM electro-optical sight. This increases the 'gun envelope' from 1120x230x88mm (PSO-1 sight) to 1220x271x118mm, its weight rising from 4.3 to 5.9kg. Sighting distance is 300m.

SVDN3 The N3 version has a large-body NSPU-3 sight, which is shorter and appreciably more compact than the elongated NSPUM type. The 'gun envelope' measures 1220x303x140mm, the weight of the gun/sight combination being approximately 5.8kg. Like the NSPUM, the NSPU-3 is sighted for distances up to 300m.

USSR

SVD rifles have been used wherever Soviet influence was strong. Captured guns were used by the warring factions in Afghanistan, while others have appeared in Iran. Production has been undertaken in Bulgaria, the People's Republic of China, Egypt, Hungary and Poland. Most of the optical sights have been made in the Soviet Union and the German Democratic Republic, but it is probable that the Chinese also make large numbers of them. Markings usually determine the country of origin.

The SVD is very light by modern sniper rifle standards, and it has an unsophisticated trigger. However, it has always been seen as an integral part of infantry equipment — a tradition in the Soviet Army dating from the early 1930s — and its drawbacks have simply been accepted in return for large-scale distribution.

SVD sniper rifle

Snayperskaya Vintovka Dragunova (SVD)
Made in the Izhevsk ordnance factory.

Specification
Calibre 7.62mm (.30)

114

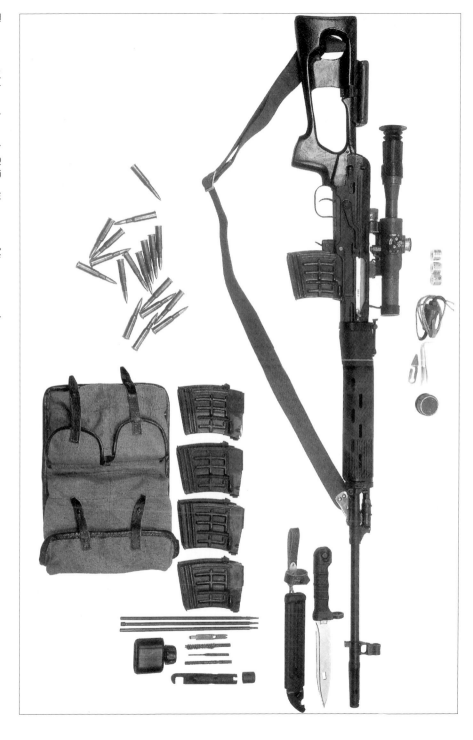

116 *An Afghan mujehaddin with a captured SVD, giving scale to the light and slender barrel.*

Cartridge 7.62x54 Soviet rimmed
Operation Gas-operated, semi-automatic
Length 1220mm
Weight 4.3kg, with empty magazine and PSO-1 sight
Barrel 545mm, four grooves, right-hand twist
Magazine Detachable ten-round box
Rate of fire N/A
Selector markings 'O' above 'Π'
Muzzle velocity 830m/sec
Bayonet Detachable knife pattern

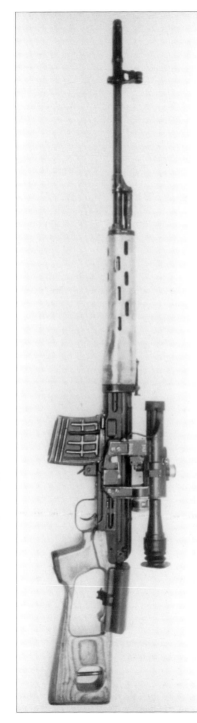

A Soviet SVD, demonstrating the method of mounting the PSO-1 sight on the left side of the receiver.

The Dragunov was accepted for service in 1963, the first guns entering service in 1966. Credited to a design team led by Evgeniy Dragunov and Ivan Samoylov, the SVD shares the action of the standard gas-operated Kalashnikov, with lugs on the bolt head turning into recesses in the receiver wall. It is equipped with a short-stroke gas piston adapted from the pre-war Tokarev.

Owing to the cutaway butt and combined pistol grip, the Dragunov is readily identifiable. The Romanian FPK and one of the Yugoslavian sniper rifles are superficially similar, but closer inspection reveals that both are built on the standard Kalashnikov action.

The slender SVD barrel has a three-slot muzzle-brake/compensator and a two-position gas-port, which can be adjusted with the aid of a cartridge-case rim. The trigger is based on the AK, but the combined safety lever/selector restricts it to single shots. The standard 4x24 ⊳SO-1 optical sight, which clamps to a 'rail on the lower left side of the receiver, is graduated to 1300m and has a passive infra-red detection capability. The long rubber eye cup is most distinctive, and the reticle can be illuminated by power from a small battery carried in the integral mount.

Ribs pressed into the sides of the box magazine improve the feed of the clumsy rimmed cartridges. Soviet publications have reported that development of the magazine's body was a particularly lengthy process.

A typical example will display the maker's mark (the Izhevsk arrow-in-triangle) under the receiver, ahead of the

117

magazine. It will be above the date and the serial number, which will be repeated on the bolt, on the fore-end ahead of the receiver, and on the sight mount rail.

The front of the optical-sight body will be marked with the designation 'PCO-1' above '☆', a hammer and sickle and two outlined blocks, which may be part of the manufacturer's trademark. These are followed by a serial number above the date of manufacture. The battery control switch is marked 'BKЛ' and 'BЫK'.

Yugoslavia/Serbia

7.62mm Model 76 sniper rifle
Poluautomatska puska vz. 76

Made by Zavodi Crvena Zastava, Kragujevac.

Specification
Calibre 7.62mm (.30)
Cartridge 7.62x54 rimmed
Operation Gas-operated, selective fire
Length 1135mm
Weight 4.25kg, without magazine or sight
Barrel 550mm, four grooves, right-hand twist
Magazine Detachable ten-round box
Rate of fire N/A
Selector markings 'U' above 'J'
Muzzle velocity 850m/sec
Bayonet Optional

Based on the standard Kalashnikov action, this sniping weapon has a long barrel and a cutaway butt with an integral pistol grip — inspired by the Dragunov (SVD) rifle, which has not been made in Yugoslavia. A mount on the receiver accepts an indigenous copy of the Soviet PSO-1 optical sight (known as ON3 or M75), together with the standard NSP-2 and NSP-3 night-vision types.

Similar guns
Rifles of this type may be encountered with the standard Kalashnikov-type butt and pistol grip instead of the SVD cutaway type.

7.92x57

Yugoslavia/Serbia

7.92mm Model 76 sniper rifle
Made by Zavodi Crvena Zastava, Kragujevač.

Specification
Calibre 7.92mm (.311)
Cartridge 7.92x57 Mauser rimless
Operation Gas-operated, selective fire
Length 1135mm
Weight 4.5kg, without magazine or sight
Barrel 550mm, four grooves, right-hand twist
Magazine Detachable ten-round box
Rate of fire N/A
Selector markings 'U' above 'J'
Muzzle velocity 850m/sec
Bayonet Optional

Adopted as the sniper rifle of the Yugoslavian armed forces, this is little more than a standard Kalashnikov action enlarged to handle full-power cartridges. The straight-comb butt with vestigial pistol grip, the ventilated fore-end and the substantial pistol grip are all wooden. Optical or electro-optical sights may be attached to a bracket on the left side of the receiver, and a bipod may be fitted. Barrel length and the shallow magazine distinguish the gun instantly. Powerful and accurate, it is let down by its coarse, military-pattern trigger.

The Yugoslavian 7.92x57 M76 sniper rifle fitted with an ON-3 optical sight, a copy of the Soviet PSO-1.

8.2x53R

USSR

Medved sporting rifle
Made by the Izhevsk ordnance factory.

Specification
Calibre 8.2mm (.323)
Cartridge 8.2x53 rimmed
Operation Gas-operated, semi-automatic
Length 1110mm
Weight 3.75kg, empty
Barrel 550mm, four grooves, right-hand twist
Magazine Integral five-round box
Rate of fire N/A
Selector markings N/A
Muzzle velocity 775m/sec
Bayonet None

One of the most interesting sporting rifles made in the Soviet Union, the Medved (Bear) shares the gas system and rotating-bolt lock of the SVD. It features a refined receiver design and a conventional straight-comb butt. Chequering appears on the pistol grip and the fore-end.

Introduced in 1963, Medved is equipped with military-style open sights and is fitted with a cleaning rod beneath the barrel. A stubby 4x optical sight is carried in a trapezoidal mount that attaches to the lower left-hand side of the receiver, above the magazine. Most examples of this gun are chambered to accommodate adaptations of the standard Mosin-Nagant rifle cartridge.

Production of the Medved, possibly never very large, had come to an end by the mid-1970s.

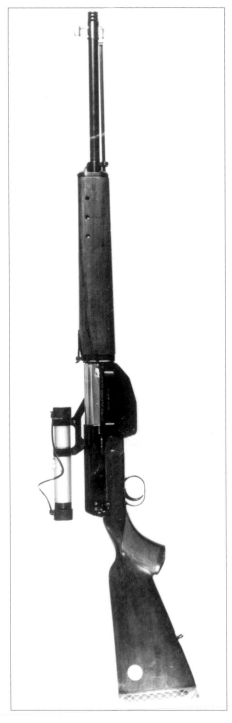

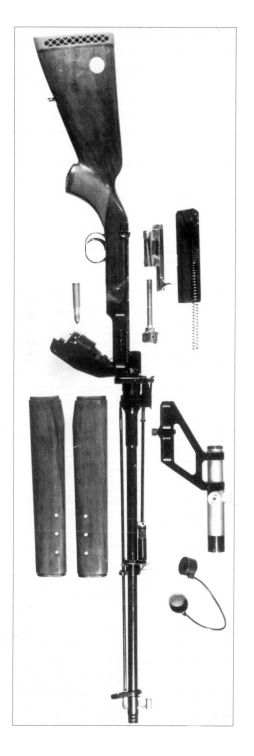

The Medved rifle dismantled into its main components. Note the shape of the telescope-sight mount.

9.2x53R

USSR

Medved sporting rifle

Made by the Izhevsk ordnance factory.

Specification

Generally similar to 8.2mm version (q.v.), except chambering and muzzle velocity (710 m/sec)

This version of the Medved rifle is chambered for a larger sporting-rifle cartridge, apparently developed by Sako from the standard Russian rimmed service-rifle cartridge.

Shotguns

Russia

Sayga-410

Samozaryadnyi gladkostvolnyi karabin Sayga-410

Made by Izhmash AO, Izhevsk, Udmurt Republic.

Specification

Calibre 10.4mm (.410)

Cartridge .410 shotgun, 3in/76mm case

Operation Gas-operated, semi-automatic

Length 1110 or 1170mm

Weight 3.4kg, with magazine

Barrel 518 or 610mm, smoothbore

Magazine Detachable box: two or four or ten rounds

Rate of fire N/A

Selector markings 'S' above 'F'

Muzzle velocity N/A

Bayonet None

Introduced in 1994, the Sayga-410 is a straightforward smooth-bore modification of the Kalashnikov assault rifle, outwardly resembling the Finnish Petra sporting rifle (q.v.). Many of the parts are identical (eg the receiver, the bolt cover and the selector lever), even though changes have been made to the bolt and bolt carrier to allow

large-diameter shotgun cartridges to be used. The rate-reducing mechanism and the automatic sear are omitted, as the gun is restricted to semi-automatic operation. The magazine of the .410 gun is raked to the rear, unlike the otherwise comparable 20- and 12-bore versions.

Sights are restricted to a fixed blade and an adjustable U-notch on a rib, or strap, above the gas-piston tube. A 3.5x PO optical sight can be mounted to the bracket on the lower left side of the receiver when required.

The standard Sayga-410 has a black plastic pistol-grip butt with a straight

A typical Sayga semi-automatic shotgun.

comb, and a sculpted fore-end of the same material. Swivels are provided on the underside of the butt and fore-end. A deluxe version has wooden furniture, with a Monte Carlo comb on the butt and thin white plastic spacers accompanying the butt plate and the pistol-grip cap. Alternatively, the butt can be replaced with a wooden 'revolver' hand grip.

The barrels may be slightly choked or cylinder-bored, although Paradox half- and full-choke attachments can be supplied with the latter. Standard equipment includes a spare magazine, an oiler and a selection of cleaning accessories in a plastic container.

Sayga-410K
Samozaryadnyi gladkostvolnyi karabin Sayga-410K
Made by Izhmash AO, Izhevsk, Udmurt Republic.

Specification
Generally as Sayga-410, except:
Length 835mm (butt extended), 595mm (butt folded)
Weight 3.3kg, with magazine
Barrel 330mm, smoothbore

This shotgun has a short barrel and a straight-wrist, folding butt moulded in black plastic. The plastic pistol grip is fitted beneath the receiver in the manner of the assault rifles, which means that it is considerably closer to the magazine than that of the standard Sayga-410. A safety feature, which locks the bolt and the trigger, ensures that the gun cannot be fired with the butt in the forward position; the lock is released simply by extending the stock.

The swivels are fitted beneath the plastic fore-end and on the right side of the butt wrist. A longitudinal groove in the butt serves as an additional identifying feature.

Sayga-410S
Samozaryadnyi gladkostvolnyi karabin Sayga-410C
Made by Izhmash AO, Izhevsk, Udmurt Republic.

Specification
Generally as Sayga-410, except:
Length 1080mm (butt extended), 835mm (butt folded)
Weight 3.3kg, with magazine
Barrel 570mm, smoothbore

Another variation of the basic design, the 410S is little more than a long-barrelled 410K, with the same folding butt and safety features.

Sayga-20
Samozaryadnyi gladkostvolnyi karabin Sayga-20
Made by Izhmash AO, Izhevsk, Udmurt Republic.

Specification
Calibre 15.6mm (20-bore, .616)
Cartridge 20-bore shotgun, 2.75in/70mm or 3in/76mm case
Operation Gas-operated, semi-automatic
Length 1135mm
Weight 3.2kg, with magazine
Barrel 570mm, smoothbore
Magazine Detachable box: two, five or eight rounds
Rate of fire N/A
Selector markings 'S' above 'F'
Muzzle velocity 330m/sec, with Baikal ammunition
Bayonet None

Made only in small numbers during 1994-6, this shotgun was offered as an alternative to the Sayga-410. It has an adjustable gas regulator, marked '1' for standard 70mm-case and '2' for magnum 76mm-case ammunition. Capable of surprising accuracy — including groups of 50-60mm at 50m with Brenneke-type slugs — the Sayga-20 was briefly touted as a personal protection weapon, but the popularity of 12-bore rivals prevailed and the Sayga-12 (q.v.) was substituted.

The Sayga-20 is similar externally to the Sayga-410, although the barrel is thicker and the magazine, which is noticeably larger, is raked forward instead of to the rear.

Sayga-20K

Samozaryadnyi gladkostvolnyi karabin Sayga-20K

Made by Izhmash AO, Izhevsk, Udmurt Republic.

Specification
Generally as Sayga-20, except:
Length 910mm (butt extended), 670mm (butt folded)
Weight 3.1kg, with magazine
Barrel 430mm, smoothbore

This is a short-barrel version of the 410S, with a folding butt and an assault rifle-style pistol grip. The swivels are fitted to the right-hand side of the butt wrist and the underside of the fore-end. Furniture is synthetic.

Sayga-20S

Samozaryadnyi gladkostvolnyi karabin Sayga-20C

Made by Izhmash AO, Izhevsk, Udmurt Republic.

Specification
Generally as Sayga-20, except:
Length 1050mm (butt extended), 810mm (butt folded)
Weight 3.2kg, with magazine
Barrel 570mm, smoothbore

This is simply a long-barrelled version of the Sayga-20K (q.v.), retaining the folding plastic butt and the automatic safety feature. The swivels are fitted to the butt wrist and fore-end.

Sayga-12

Samozaryadnyi gladkostvolnyi karabin Sayga-12

Made by Izhmash AO, Izhevsk, Udmurt Republic.

Specification
Calibre 18.2mm (.729)
Cartridge 12-bore shotgun: 70, 73 or 76mm case
Operation Gas-operated, semi-automatic
Length 1145mm (standard butt, 930mm pistol butt)
Weight 3.6kg, with magazine
Barrel 675mm, smoothbore
Magazine Detachable box: two or six rounds
Rate of fire N/A
Selector markings 'S' above 'F'
Muzzle velocity N/A
Bayonet None

Basically, this auto-loading shotgun is a heavier version of the Sayga-20, adapted to handle the larger-diameter, but much more popular, 12-bore cartridge. The gun is easily identified by its large barrel and the length of its straight-sided plastic magazine, which is raked forward. The gas regulator can be adjusted for stand-ard ('1') or magnum ('2') ammunition.

Guns can be obtained with either a full choke or an adjustable fixture that gives a choice of full choke, half choke or cylinder bore. The trigger-pull can be adjusted from 1.5kgf to about 3.7kgf.

Most guns are fitted with plastic furn-ture, which includes a straight-combed pistol grip butt. However, a wooden butt, with a Monte Carlo comb and white spacers, and a 'revolver-type' hand grip are available as optional fittings (see Sayga-410).

Sayga-12K

Samozaryadnyi gladkostvolnyi karabin Sayga-12K

Made by Izhmash AO, Izhevsk, Udmurt Republic.

Specification
Generally similar to Sayga-12, except:
Length 910mm (butt extended), 670mm (butt folded)
Weight 3.5kg, with magazine
Barrel 430mm, smoothbore

A minor variant of the Sayga-12, this shotgun features a short barrel, a butt that can be folded along the left-hand side of the receiver, and an assault rifle-type pistol grip. A safety interlock prevents the gun from being fired when the butt is folded forward. The rear sling swivel is attached to the right-hand side of the butt wrist.

Sayga-12S

Samozaryadnyi gladkostvolnyi karabin
Sayga-12C
Made by Izhmash AO, Izhevsk,
Udmurt Republic.

Specification

Generally as Sayga-12, except:
Length 1060mm (butt extended),
820mm (butt folded)
Weight 3.6kg, with magazine
Barrel 570mm, smoothbore

Another version of the basic Kalashnikov
12-bore auto-loading shotgun, the Sayga-
12S is distinguished from the 12K by the
length of its barrel. It shares the folding
butt, pistol grip, safety interlock and sling-
swivel positions of the short-barrel gun.

ACCESSORIES

Ammunition

The Russian 'intermediate' 7.62x39mm cartridge, known as the obr. 1943g (M43 in Western parlance), was inspired by the German 7.9mm Kurz type. Developed at the end of the Second World War, it did not reach quantity production until after 1946. The standard Russian round has a rimless, bottle-necked, copper-washed steel case, is Berdan primed and is 38.6mm long. However, some made in Czechoslovakia and the German Democratic Republic are lacquered steel, while brass-case rounds have been made in Finland and the UAR.

PS The standard boat-tailed bullet, with a single cannelure and a steel (*stalnoi*) core, weighs 7.9gm and develops a muzzle velocity of about 710m/sec from a 1.6gm charge. There are no notable distinguishing marks.

BZ The standard 7.8gm 'intermediate' armour-piercing/incendiary bullet has a plain boat-tail and a black tip above a red band. Its steel core lies immediately ahead of a small pellet of incendiary compound, which is ignited when the bullet strikes a hard surface.

T-45 The flat-based 7.4gm tracer bullet is distinguished by a green tip. The trace pellet lies in the base of the bullet, behind the lead-alloy core.

Z The standard 6.6gm incendiary/ranging

A Swedish soldier fires an FFV 915 rifle grenade from the 5.56mm FFV 890C (modified Galil) rifle.

bullet has a red tip. It has a central steel core, ahead of a trace pellet, with an incendiary compound in the nose cavity, between the core and the envelope wall.

The low-impulse 5.45mm assault rifle cartridge — loaded with a standard steel-core bullet or a tracer pellet — was designed by a group headed by engineer Viktor Sabelnikov and including Lidiya Bulavskaya, Boris Semin, Mikhail Fedorov, Petr Sazonov and Petr Korolev. It has been chambered in the AK-74 assault rifle and the RPK-74 light machine-gun, and developed as a counter to the US-inspired 5.56mm pattern by way of a 5.6x39mm Soviet sporting-rifle cartridge that dates back to the 1960s.

Normally, the bottle-necked cartridge case is made of lacquered steel. Loaded rounds measure 57mm overall, the rimless cases being 39.6mm long, and the bullets 25.5mm.

PS This bullet is an interesting design, having a relatively high length-to-diameter ratio (4.53 calibres) and a complex core within a conventional cupro-nickelled steel envelope. The core consists of a steel after-core and a short lead plug with an air-gap in the nose. The tip crushes when it strikes flesh or wood, causing the bullet to tumble and become much more destructive than otherwise it would be. However, this effect is achieved at the expense of considerable manufacturing complexity. The PS weighs about 3.44gm and is boat-tailed. The propellant is a double-base pattern that generates a muzzle velocity of about 900m/sec.

Others Armour-piercing and tracer bullets have been reported, but their designations remain uncertain. The former have black tips (possibly above a red ring, which indicates an incendiary capability), the latter, green tips.

Bayonets

The bayonet for the AK assault rifle (*Shtik-nozh za Avtomata Kalashnikova*) is a curious design, giving rise to the mistaken suggestion that it was developed after the rifle had been introduced to meet an unexpected need.

The basic design is conventional, having a 210mm double-edged blade with typical Russian-style fullers, but the attachment system relies on a 17.7mm-diameter muzzle ring on the crossguard (retained by rivets or a brazed joint) and a

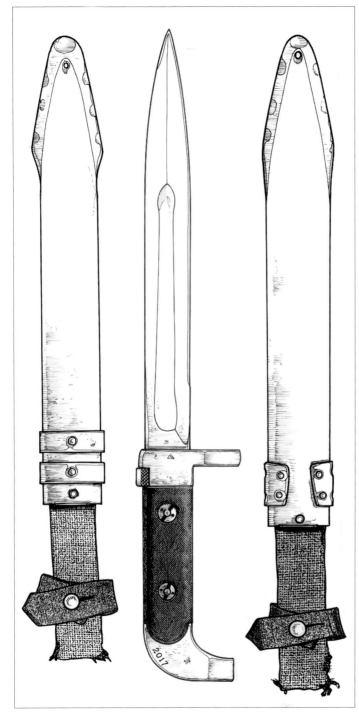

The knife bayonet and scabbards for the AK-47 rifle. Drawings by John Walter.

131

Evolution of the Chinese Kalashnikov bayonet. At first, the Soviet SKS, which has a knife-type blade (top), was copied in China. Later, this was replaced by a longer triangular-section spike (centre). A shorter version of this spike is used on the Type 56 Kalashnikov (bottom).

The Chinese Type 56 spike bayonet. The catch sleeve can be slid forward against the pressure of a coil spring, which is anchored against the blade-tang shoulder. Drawing by John Walter.

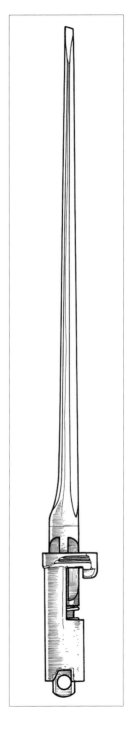

semi-ring formed by an upward extension of the pommel. The latter is slipped over the barrel and pushed back over the gas-port housing.

As the true muzzle ring locks over the tip of the muzzle, a double-prong lock (controlled by a sprung latch under the hilt) springs up around the cleaning-rod housing, under the front sight. No variations of the basic bayonet have been seen, apart from differences in grip material. However, there are several different scabbards (*nozhik*), which have webbing belt-loops held by either one wide or two narrow encircling bands.

The hilt, with its unusual pommel 'ears', has a recessed backstrap to clear the cleaning rod and is most uncomfortable to hold. In general, the grips are of brown-red plastic or plastic-impregnated wood fibre, being retained by two bolts and slotted plain-head nuts.

The earliest AKM-type tool bayonet retains the muzzle-ring and pommel-extension attachment method, but handling is improved by the use of two well-rounded, plastic-impregnated wood-fibre grip plates. These are riveted together, although the rivet heads are covered with wood-fibre discs. A shrouded press-stud in the pommel controls attachment to the rifle.

The bowie-point blade has a short serrated section on the under-edge, while a hole in the centre of the blade receives a lug on the tip of the scabbard. This allows the bayonet and scabbard to act as a scissors-type wire cutter.

The scabbard itself consists of a plastic/wood-fibre body with a short sleeve-type insulator. A short strap is attached to the back of the scabbard, anchoring the hilt-retaining loop and the spring-clip for the belt.

An improved tool bayonet, made in huge numbers in several countries, has an inverted blade and a much more angular grip. This is held to the tang by a single bolt, and there is a separate pommel rather than one buried in the wood fibre.

The new scabbard, although similar to the original, features an improved parallel-sided body and a modified suspension system that makes use of a belt loop rather than a spring-clip. A leather thong can be looped through a transverse hole in the pommel and around a hook on the vestigial quillon.

The most recent tool bayonet developed for the Kalashnikov shares the inverted blade of its predecessor, but has a rounded, tube-like plastic grip with several circumferential ribs to improve grip. The shrouded press-stud is retained, together with the insulated scabbard.

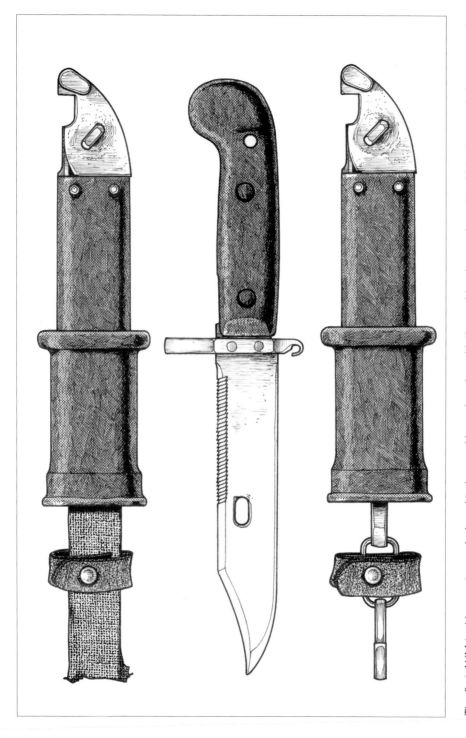

The first AKM tool bayonet was designed to be used in conjunction with the scabbard as a wire cutter. Note the rounded contours of its hilt. The earliest scabbard has a webbing belt loop, but later types have D-ring-and-loop fittings. Drawings by John Walter.

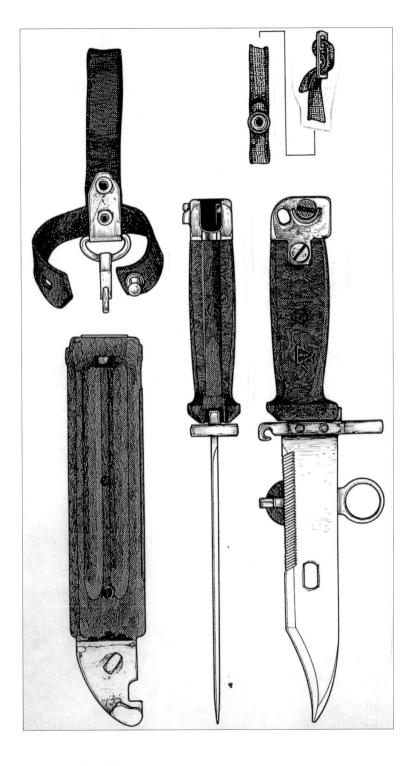

The second-pattern AKM tool bayonet has squared contours and an inverted saw-back blade. This Izhevsk-made example also displays a shrouded press-stud in the pommel. Drawings by John Walter.

136 *The current Izhevsk-made tool bayonet for the Kalashnikov rifle.*

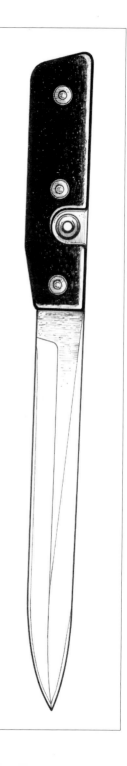

Bayonets used in Finland, made by Fiskars and based on a traditional hunting-knife design, are quite different from their Soviet and Russian counterparts. The original pattern has a plain wooden grip, but the perfected m/62 has ribbed plastic. Drawing by John Walter.

Sights

Some Kalashnikov rifles can be fitted with optical or electro-optical sights. Guns will also be found with a curious night sight, comprising a luminous slider on rails, which can be clipped to the aperture in the front-sight mounting block and used in conjunction with two dots on an auxiliary back-sight notch plate.

The original Kalashnikov sights consisted of an adjustable pin at the front and an adjustable leaf-and-slider at the rear, acting against the shaped sides of the sight bed to provide the appropriate elevation settings. Most 7.62mm-calibre AK-type guns have back-sight leaves graduated to 800m, though the numerals may often be encountered in Arabic. Later 7.62mm AKM and the 5.45mm guns customarily have 1000m sights, though the construction is practically identical. The back-sight notches of most of the light machine-gun adaptations can be moved laterally by turning a finger wheel on the right side of the sight block.

The AK and AKM, like all assault rifles of their type, are intended for use at comparatively short ranges. Consequently, until recent years, few attempts have been made to improve their sights other than to add Tritium dots to improve performance in dusk or poor light. The need for long-range accuracy has always

A Soviet NSP-2 infra-red sight mounted on a Czechoslovakian vz. 58 assault rifle. First-generation sights can be very cumbersome.

The Soviet PSO-1 optical sight. This particular example lacks the rubber sight hood that normally projects to the rear, over the butt.

(Above and right) The Izhevsk-made SVD (Dragunov), shown with the current generation of electro-optical sights: NSPUM (above) and NSPU-3 (right).

been provided, in the Soviet and Russian armies at least, by a combination of the Dragunov rifle (SVD, q.v.) and its PSO-1 optical sight.

With the advent of compact infra-red and the electro-optical equipment in the last twenty years, however, the situation has changed appreciably. Guns are now fitted with a rail on the lower left side of the receiver to accept a mounting bracket for low-light and image-intensifying sights.

Russian equipment, which includes the current designs shown on these pages, is sturdy and dependable. However, its performance still falls some way short of the latest US and Western European designs.

A Yugoslavian M76 sniper rifle fitted with an image-intensifying sight.

Silencers

The AKM and AK-74 can be adapted by replacing the muzzle-brake/compensator with a PBS-1 or comparable silencer. The basic designs are similar, except for calibre and size.

A typical PBS-1 comprises a two-part cylindrical casing, joined by a screw thread. Up to ten baffle plates, each with a central axial hole for the projectile, are carried on three longitudinal rods that act as spacers. Behind the baffle-plate assembly is a 20mm-thick rubber plug, which in turn is positioned ahead of a short conical void. From this void radial holes communicate with an annular expansion chamber.

When the gun is fired, the bullet traverses the void, allowing some gas to approach the expansion chamber, and bleed into the expansion chamber, and pierces the rubber plug. The rubber tends to spring back behind the projectile, and inevitably the pressure of the propellant gas is reduced. As the gas escapes forward, past the rubber plug, it loses momentum among the baffles before being vented at the muzzle.

Silencers require special cartridges with heavy bullets and reduced powder charges. However, after several shots, the hole in the rubber plug will begin to approach the diameter of the projectile, reducing efficiency until eventually the insert must be replaced.

Virtually all silencers are constructed in a similar manner and operate on the same principles, but a radically different approach is to use special noiseless cartridges, like those of the PSS pistol. In this respect, the 7.62mm SP-3 noiseless cartridge has been chambered in a shortened, silenced AK-74UB rifle.